EYE AND VISION RESEARCH DEVELOPMENTS

AGE-RELATED MACULAR DEGENERATION

PREVALENCE, RISK FACTORS AND CLINICAL MANAGEMENT

EYE AND VISION RESEARCH DEVELOPMENTS

Additional books in this series can be found on Nova's website under the Series tab.

Additional e-books in this series can be found on Nova's website under the e-book tab.

EYE AND VISION RESEARCH DEVELOPMENTS

AGE-RELATED MACULAR DEGENERATION

PREVALENCE, RISK FACTORS AND CLINICAL MANAGEMENT

MELANIE CARTER
EDITOR

NOVA BIOMEDICAL
New York

Library of Congress Cataloging-in-Publication Data

ISBN: 978-1-63483-329-5

Library of Congress Control Number: 2015944484

Published by Nova Science Publishers, Inc. † New York

Contents

Preface

Age-related macular degeneration (AMD) is the leading cause of blindness in individuals older than 65 years of age. The authors of this book discuss the role of genetics, sunlight, diet, cardiovascular factors, smoking and alcohol as possible risk factors for AMD. Furthermore, the dry form is more common and accounts for about 85-90% of AMD patients in the USA, while Japanese AMD patients predominantly progress to wet-form or polypoidal choroidal vasculopathy (PCV). The function of a serine protein gene, HTRA1, in wet-form AMD patients is explored in detail. Conversely, the final chapter evaluates the therapeutic effects of high omega-3 fatty acids as anti-inflammatory agents in dry AMD patients. The preliminary findings indicate a promising therapeutic regime for dry AMD and perhaps for other retinopathies as well.

Chapter 1 – Age related macular degeneration (AMD) is the leading cause of blindness in individuals older than 65 years of age. It is a multifactorial disorder and identification of risk factors enables individuals to make life style choices that may reduce the risk of disease. This review discusses the role of genetics, sunlight, diet, cardiovascular factors, smoking, and alcohol as possible risk factors for AMD. Genetics plays a more significant role in AMD than previously thought, especially in younger patients, histocompatibility locus antigen (HLA) and complement system genes being the most significant. Whether the risk of AMD is increased by exposure to sunlight, cardiovascular risk factors, and diet is more controversial. Smoking is the risk factor most consistently associated with AMD. Current smokers are exposed to a two to three times higher risk of AMD than non-smokers and the risk increases with intensity of smoking. Moderate alcohol consumption is unlikely to increase the risk of AMD. Optometrists as front-line informers and educators of ocular health play a significant role in increasing public awareness of the risks of AMD. Cessation of smoking, the use of eye protection in high light conditions, dietary changes, and regular use of dietary supplements should all be considered to reduce the lifetime risk of AMD.

Chapter 2 – Age-related macular degeneration (AMD) is a leading cause of vision loss and blindness in the elderly. The dry form is more common and accounts for about 85~90% of AMD patients in US, while Japanese AMD patients predominantly progress to wet-form or polypoidal choroidal vasculopathy (PCV). Recent studies have shown *HTRA1*, a serine protease gene, as major risk factor for wet form AMD (De Wan et al., Science 2006). Furthermore, the authors reported that the Japanese typical wet form AMD patients showed significant association with *ARMS2/HTRA1* (Goto, Akahori et al., JOBDI 2009). The purpose of this study is to elucidate the function of *ARMS2/HTRA1* gene promoter in wet-form AMD patients. The promoter sequence experiment showed that a great number of AMD patients had specific indel mutation in 3.8 kb upstream of *HTRA1* gene. 2~3-fold increase of promoter activity was observed in indel *HTRA1* promoter compared to control sequence (Iejima et al., JBC 2015). Furthermore, the authors created transgenic mice ubiquitously overexpressing mouse *HtrA1* using the chicken act in promoter, continuous induction of *HtrA1 in vivo* was shown to lead to CNV, similar to wet AMD patients (Nakayama, Iejima et al., IOVS 2014). These results suggest that human *HTRA1* expression is enhanced by AMD specific indel mutation in the promoter region of *HTRA1* gene, and this enhanced *HTRA1* may be concerned with induce retinal neovasucularization.

Chapter 3 – One of the major causes of reduced vision over the age of 50 is age-related macular degeneration (AMD). Although such a common pathology, there are no current guidelines for the first-line treatment of dry AMD. The aim of this study is to evaluate the therapeutic effects of high omega-3 fatty acids as anti-inflammatory agents in two sub-groups of dry AMD patients with 1) mild to moderate visual impairment and 2) with severe visual impairment (blindness). The key feature of this investigation is the frequent monitoring of the levels of specific fatty acids in patient's blood in order to adjust the treatment dose within the ideal therapeutic window. Following the positive outcome from the authors' initial observational studies in patients with dry AMD, who demonstrated significant improvement in visual acuity (gain of ≥ 1 line of vision in 4.5 months) when taking a total of 5 g/day eicosapentaenoic acid (EPA) and docosahexaenoic acid (DHA), additional studies were encouraged. The latest data which is presented in this chapter suggests that the eyes which had the greatest gain in vision (≥ 15 letters gain at 6 months) were from patients with mild to moderate visual impairment, who were taking between 5-7.5 g/day EPA and DHA and had a ratio of arachidonic acid (AA)/EPA < 2. In addition, a sub-group of dry AMD patients with severe visual impairment (< 6/60), showed significant increase in their visual acuity only 3 months following treatment with omega-3 fatty acids. The preliminary results indicate a promising therapeutic regime for dry AMD and perhaps for other types of retinopathies as well. Although initial results are encouraging, further investigations are necessary to establish a better understanding of the mode of action of these supplements and to observe their long-term effects.

In: Age-Related Macular Degeneration
Editor: Melanie Carter
ISBN: 978-1-63483-329-5

Chapter I

Overview of Risk Factors for Age-Related Macular Degeneration (AMD)

Richard A. Armstrong [*] ***and Maryam Mousavi***
Vision Sciences, Aston University, Birmingham, UK
Corresponding author: Dr R.A. Armstrong, Vision Sciences, Aston University, Birmingham, UK

Abstract

Age related macular degeneration (AMD) is the leading cause of blindness in individuals older than 65 years of age. It is a multifactorial disorder and identification of risk factors enables individuals to make life style choices that may reduce the risk of disease. This review discusses the role of genetics, sunlight, diet, cardiovascular factors, smoking, and alcohol as possible risk factors for AMD. Genetics plays a more significant role in AMD than previously thought, especially in younger patients, histocompatibility locus antigen (HLA) and complement system genes being the most significant. Whether the risk of AMD is increased by exposure to sunlight, cardiovascular risk factors, and diet is more controversial. Smoking is the risk factor most consistently associated with AMD. Current smokers are exposed to a two to three times higher

[*] Email: R.A.Armstrong@aston.ac.uk. Tel: 0121-204-4102; Fax: 0121-204-4048.

risk of AMD than non-smokers and the risk increases with intensity of smoking. Moderate alcohol consumption is unlikely to increase the risk of AMD. Optometrists as front-line informers and educators of ocular health play a significant role in increasing public awareness of the risks of AMD. Cessation of smoking, the use of eye protection in high light conditions, dietary changes, and regular use of dietary supplements should all be considered to reduce the lifetime risk of AMD.

Keywords: age-related macular degeneration (AMD), risk factors, genes, smoking, sunlight, diet, epidemiology

Introduction

Age-related macular degeneration (AMD) is the most important cause of blindness in industrialised countries in individuals over 65 years of age. In the UK, for example, overall prevalence of late stage AMD is 2.4% of the population (Owen et al., 2012) and approximately 10% of patients 66 - 74 years of age will exhibit evidence of macular degeneration, prevalence increasing to 30% in patients 75 - 85 years of age. As a consequence, there are estimated to be 513,000 individuals currently in the UK with the visual impairment characteristic of AMD suitable for registration as seriously visually impaired. This figure is predicted to rise to 679,000 by 2020 (Owen et al., 2012) as the proportional of more elderly individuals in the population continues to increase, a phenomenon likely to be repeated in many countries.

There are two forms of AMD, viz., the 'atrophic' or 'dry' form and the 'exudative' or 'wet' form. The dry form is characterized by degeneration of the macular pigment epithelium, choriocapillaris, and photoreceptor cells. Larger areas of degeneration result from the merging of small areas of atrophy and this state is often referred to as 'geographic atrophy.' By contrast, in the 'exudative form,' new choroidal vessels are formed ('neovascularisation') and ultimately, retinal detachment, oedema, and haemorrhage may occur resulting in degeneration of the macula. Age-related maculopathy (ARM) is a term often used to describe the early stages of AMD. It is most apparent clinically after 50 year of age and is characterized by discrete drusen and increased deposition of retinal pigment.

The causes of AMD are not well understood but many factors appear to be involved: aging, genetics, smoking, diet, alcohol, cardiovascular function, and exposure to sunlight all having been implicated (Yoshida et al., 2000; Delcourt et al., 1998, 2001). The main objectives of this review are: (1) to review the major risk factors associated with AMD, (2) to attempt to rank these factors in order of significance, and (3) to suggest modifications in life-style that may reduce the lifetime risk of AMD.

Prevalence

In the UK, overall prevalence of late stage AMD is 2.4% and annual incidence per 1000 individuals of late AMD, dry AMD, and wet AMD is 4.1, 2.4, and 2.3 respectively in females and 2.6, 1.7, and 1.4 respectively in males (Owen et al., 2012). Ethnic differences in the prevalence of AMD have also been reported. Hence, the risk of AMD varies among Americans of different Asian origin (Stein et al., 2011). Overall, 5.4% of Asian Americans have dry and 0.49% wet AMD, Chinese and Pakistani Americans having a significantly increased risk of dry AMD compared with non-hispanic whites. By contrast, Japanese Americans have a 29% decreased risk of dry AMD (Stein et al., 2011). There were no significant differences in the risk of wet AMD in Asian Americans of any ethnicity compared with white Americans. In a further study of ethnicity in the US (van der Beek et al., 2011), overall prevalence was 5% for dry AMD and 0.76% for wet AMD. Blacks have a reduced risk of dry and wet AMD at ages 60 and 80 years relative to whites while Americans of hispanic origin have an 18% reduced risk. By contrast, Asian Americans have a 28% increased risk of dry AMD but a 46% reduced risk of wet AMD at age 80. A particularly high prevalence of AMD has been recorded in an Oklahoma Indian population (the 'Strong Heart Study'), overall prevalence being 35.2% and prevalence of late AMD being 0.81% (Butt et al., 2011). Hence, both the prevalence of AMD and the development of wet from dry AMD appear to vary significantly with ethnicity.

Outside USA, a Japanese study at Funagata (Kawasaki et al., 2008) found that the prevalence of early AMD in individuals greater than 35 years of age was 3.5%, and of late AMD 0.5%, while the age-standardised prevalence in the right eye was 4.1%. Meta-analysis has been used to model future rates of AMD in Switzerland (Bauer et al., 2008), late AMD being predicted to rise from 37,200 in 2005 to 52,500 in 2020, and 93,200 in 2050. In rural China, prevalence of wet and dry AMD is 1.45% and 1.55% respectively, increasing with age (Bai et al., 2008). An overview of epidemiological studies from around the world (Klein, 2007) suggests prevalence of early AMD is similar in the white, black, and hispanic populations, but that whites have a greater prevalence of late AMD. Hence, prevalence studies provide considerable evidence that genetic, environmental, and lifestyle differences among populations are likely to be involved in determining the risk of AMD.

Genetics

The lifetime risk of developing late AMD is 50% for individuals that have a relative with the disease compared with 12% in families without the disease (Stine, 1989). Many genes have been implicated as possible risk factors for AMD. Most of these involve single nucleotide polymorphisms (SNP) in which one amino acid within the protein is substituted for another. These are often natural allelic variations within the population resulting in either increased risk or protection against AMD. By contrast, copy number variation (CNV), i.e., variation in the number of copies of a specific DNA sequence appears to be less important, although a small number of such associations with AMD have been recently reported (Liu et al., 2011). The major genes associated with AMD, their functions, and locations within the genome are summarised in Appendix 1.

ATP-binding Cassette Rim Protein (ABCR)

Earler studies suggested that a protein found in retinal rod cells, viz., adenosine triphosphate (ATP)-binding cassette rim protein (ABCR), may be involved in a number of ocular disorders including Stargardt's macular dystrophy, retinitis pigmentosa (RP), cone-rod dystrophy, and AMD (Shroyer et al., 1999; De la Paz et al., 1999; Souied et al., 2000; Fuse et al., 2000). The ABCR protein comprises two structurally related halves, each of which contains a section which spans the rod membrane and a section which can bind to ATP and is abundant in the outer segments of rod cells but absent from cone cells (Sun and Nathans, 1997). Within the rod outer segment, the protein is localized within the rim region of the rod membranes. ABCR may influence the active transport of a wide range of drugs, metabolites, peptides, and lipids and may also be involved in the transport of retinal derivatives, phospholipids, peptides, or other endogenous substrates across the disk membrane.

The *ABCR* gene is 100 KB long and consists of 50 exons ranging in size from 33 to 406 nucleotides (Allikmets et al., 1998). A full length transcript of the gene sequence was cloned in 1998 (Gerbers et al., 1998; Nosonkin et al., 1998) allowing the gene to be located to the short arm of chromosome 1 near to the centromere (1p21-p22.1). A large number of mutations of the *ABCR* gene have now been identified indicating that *ABCR* is an especially 'polymorphic' gene (Papaioannou et al., 2000). In five studies, however, there was little agreement concerning the importance of the *ABCR* gene in AMD. A study of dry AMD in Japan concluded that there was no relationship between allelic variation in *ABCR* and the disorder (Fuse et al., 2000) while the study of De la Paz et al. (1999) concluded that *ABCR* was not a major risk factor. By contrast, a study of wet AMD (Souied et al., 2000) concluded that 4% of familial cases of AMD may be related to *ABCR*. In addition, other studies have concluded that possession of some variants of the *ABCR* gene may increase susceptibility to the disease (Zhang et al., 1999). Recent studies have also been controversial. Hence, the presence of three significant mutations in *ABCR* was examined in eight cases of wet AMD (Shastry, 2004), the analysis failing to identify the presence of disease-causing mutations in the cases examined. However, *ABCR* mutations causing the protein to accumulate in the inner segment of rod cells have been reported to be frequent in patients with severe retinal dystrophies, including a small number of AMD cases (Wiszniewski et al., 2005).

Age-related Maculopathy Susceptibility Protein 2 (ARMS2)

Age-related maculopathy susceptibility protein 2 (ARMS2) is a cytoplasmic protein, currently of unknown function, which may play a role in various diseases of the elderly including AMD. SNP of the *ARMS2* gene are especially associated with wet AMD and the frequency of progression from dry to wet AMD (Dietzel et al., 2010) but are less associated with dry AMD (Fuse et al., 2011). In a Spanish population, *ARMS2* has been reported to be one of the principal risk factors for AMD (Brion et al., 2011).

Apolipoprotein E (APOE)

Apolipoprotein E (APOE) is a lipid transport protein essential for the catabolism of triglyceride-rich lipoproteins. It transports lipoproteins, fat-soluble vitamins, and cholesterol to the lymph system and then to the blood. *APOE* is a polymorphic gene with three major isoforms, viz., *ε2*, *ε3*, and *ε4*. These allelic forms differ only slightly but are sufficient to alter *APOE* structure and function. A pooled analysis of 15 studies (McKay et al., 2011) found associations between *APOE* alleles ε4 and ε2 with late AMD but there was no interaction with gender or smoking behavior. By contrast, Adams et al. (2012) also found an association between possession of *APOE* allele *ε2* and early onset AMD, an association that was present in non or previous smokers but not in current smokers suggesting an interaction between genetics and smoking. More recently, a meta-analysis of early onset AMD has suggested an association with *APOE* (Holliday et al., 2013).

ATP Synthase

ATP synthase is a mitochondrial enzyme important in providing energy to the cell via ATP. Mutations of the *ATP synthase* gene have been related to RP and this gene may also be implicated in AMD (Nordgaard et al., 2008).

Complement System Genes

Genes for the complement system protein factor H (*CFH*) (Nischler et al., 2011), factor B (*CFB*) and factor 3 (*C3*) (Yates et al., 2007, Maller et al., 2007) appear to be strongly associated with the risk of developing AMD (Chen et al., 2011). *CFH* is involved in the inhibition of the inflammatory response mediated via C3b by acting both as a cofactor for cleavage of C3b to its inactive form, C3bi, and by weakening the active complex that forms between C3b and factor B. C-reactive protein and polyanionic surface markers such as glycosaminoglycans normally enhance the ability of CFH to inhibit complement. Specific mutations in *CFH* (e.g., Tyr402His) reduce the affinity of CFH for C3f and may also alter the ability of CFH to recognize specific glycosaminoglycans. This change results in reduced ability of CFH to regulate the activity of complement proteins on critical surfaces, such as the retina, and may lead to increased inflammatory response at the macula. A higher risk of wet AMD is associated with polymorphisms of the *CFH* gene and with lower visual acuity after treatment (Nischler et al., 2011). A strong association between *CFH* and bilateral dry AMD has also been demonstrated (Chen et al., 2011).

DNA Excision Repair Protein (ERCC6)

The *DNA excision repair protein* gene (*ERCC6*) interacts with several transcription and excision repair proteins. Excision repair protein removes mutations resulting from UV damage resulting in the removal of a short single-stranded segment of DNA containing the mutation. A new complimentary strand is then synthesized using the remaining strand as a template. Mutations of *ERCC6* are responsible for 'Cockayne's syndrome,' a disorder characterised by photosensitivity and a pigment retinopathy (Baas et al., 2000). Some studies have also suggested an association between SNP of *ERCC6* and AMD (Baas et al., 2000) indicating that UV damage to genes which is unrepaired could be a factor in AMD.

Fibroblast Growth Factor 2 (FGF2)

Fibroblast growth factor 2 (FGF2) is a protein involved in various biological processes including mitosis, the development of blood vessels, wound healing, and tumor growth. In a Spanish population, Brion et al. (2011) has shown that although the principal genetic risk factors are *ARMS2* and *CFH*, *FGF2* may also be associated with some cases of dry AMD.

Fibulin-5

Rare forms of AMD, exhibiting an autosomal dominant pattern of inheritance, may be related to the *fibulin-5* gene. Fibulin-5 protein is an excreted extracellular matrix protein expressed in basement membranes of epithelial cells and blood vessels and is essential for the polymerization of elastin. AMD patients were screened and a statistically significant correlation observed between mutations of *fibulin-5* and AMD (Stone et al., 2004). Furthermore, point mutations were found affecting the calcium binding domains of the protein.

Hepatic Lipase (LIPC)

The *hepatic lipase (LIPC)* gene codes for the enzyme hepatic triglyceride lipase which is expressed in liver and the adrenal glands. The principle function of the enzyme is conversion of intermediate density lipoproteins (IDL) to low density lipoproteins (LDL). There may be an association between the proportion of different lipoproteins and AMD in that a specific genotype of *LIPC* has been linked with a reduced risk of both dry and wet forms of the disease (Seddon et al., 2010).

Lysyl Oxidase-like 1 (LOX1)

The *Lysyl oxidase-like 1 (LOX1)* gene codes for an enzyme involved in the development of connective tissue including that of blood vessels and has been shown to be significantly associated with wet AMD in a Japanese (Fuse et al., 2011) and a Korean population (Lee et al., 2010), although the biological foundation of this association is unclear.

Selectin-P (SELP)

Studies suggest an important role for the activity of macrophages and endothelial cell activation in the pathogenesis of AMD (Mullins et al., 2011). The *selectin* gene group function as cell adhesion molecules on the surface of activated endothelial cells and is involved specifically in the interactions between platelets and leucocytes and has been implicated in AMD. No single SNP of *SELP* is highly correlated with AMD, but a polymorphism of the gene has been found to be statistically associated with dry AMD (Mullins et al., 2011). Platelet and leucocyte interactions promote thrombus formation and are active at sites of inflammation, but it is not clear how these processes may lead to AMD.

Serpin Peptidase Inhibitor (SERPING1)

Genetic variants of the *serpin peptidase inhibitor (SERPING1)* gene, which regulates the activity of complement proteins, has recently been shown to be associated with AMD (Gibson et al., 2010). This protein inhibits the first component of the complement cascade thus regulating complement activation.

Serum Peptidase 1 (HTRA1)

The *HTRA serum peptidase 1 (HTRA1)* gene encodes for a secreted serine protease (Tong et al., 2010; Lee et al., 2010, Yang et al., 2006). The protein is a secreted enzyme proposed to regulate the availability of insulin-like growth factor (IGF) by cleaving IGF-binding protein. IGF binds to insulin-like growth receptor thus initiating intracellular signaling and subsequently, stimulating cell growth and inhibiting apoptosis (programmed cell death). Hence, *HTRA1* may inhibit cell death in the retina and mutations in this gene result in increased cell losses. Tong et al. (2010) emphasised the importance of gene interactions; *HTRA1* plays a significant role in AMD when it is found in combination with *ARMS2*.

Transferrin (Tr)

Transferrin (Tr) is an iron-binding blood plasma glycoprotein which controls the level of free iron in biological fluids (Winsper et al., 1997). Oxidative stress is regarded as a major risk factor for AMD and increased levels of reactive oxygen species (ROS) are produced by reactions catalysed by iron in the retina. Hence, changes in iron metabolism genes may be involved in AMD. Recently, SNP in the *Tr* gene have been shown to be significantly associated with AMD especially in smokers (Wysokinski et al., 2011) emphasizing the interactions that may occur between genetic susceptibility and environmental risk factors.

Voltage-Dependent Calcium Channel 3(CACNG3)

Voltage-dependent calcium channel 3 (CACNG3) is a protein subunit of the L-type calcium channel located in the cell membrane which regulates trafficking and channel gating. Recent genetic analysis has suggested there is a gene associated with AMD on chromosome 16 (16p12) (Spencer et al., 2011). At least five candidate genes occur in this region of the chromosome but the most likely to be associated with AMD is *CACNG3* (Spencer et al., 2011). These data suggest a possible link between calcium ion channel modulators and AMD but further research is necessary to establish the significance of this gene.

Hence, many genes could be potentially linked to an increased risk of AMD and the strongest case can be made for those involved in The most important genes implicated in AMD appear to be those involved in immune modulation and the complement system such as *CFH*, *CFB*, *C3*, and *SERPING1* strongly suggesting that inflammatory processed are important in AMD.

Diet and Cardiovascular Health

Several studies suggest that AMD is related to factors associated with diet and cardiovascular health and are summarized in Appendix 2. Hence, AMD has been linked to the consumption of a number of dietary constituents most notably carbohydrates and lipids. Hence, diets with a high glycemic index (dGI) are associated with a greater risk of AMD (Chiu et al., 2007a). In addition, risk of progression of the disease over eight years was significantly higher in patients characterized by a high dGI (Relative risk [RR] = 1.10, Confidence interval 1.00 - 1.20). In addition, a significant association was observed between high dGI and the presence of large drusen (Odds ratio [OR] = 1.42, 1.09 – 1.84) and a significant positive correlation was observed between dGI and the severity of AMD (Chiu et al., 2007b). The authors concluded that a 20% reduction of AMD cases could be achieved by changing to a diet significantly lower in refined carbohydrate (Chiu et al., 2007b).

Increased incidence of ARM has been observed in patients with raised C-reactive protein (Seddon et al., 2004), which is associated with increased consumption of lard and solid fat (Ngai et al., 2011). High triglyceride levels, by contrast, were associated with a lower risk and therefore, certain types of lipid could be protective against AMD. In a prospective study, association between AMD and fat intake was examined by Cho et al. (2001). Total fat intake of those patients who were the highest consumers was associated with increased risk of AMD (RR = 1.54, 1.17 – 2.01). In addition, linolenic acid consumption was associated with AMD (RR = 1.49, 1.15 – 1.94) whereas docosahexaenic acid (DHA) had a modest inverse relationship with AMD (RR = 0.70, 0.52 – 0.93). In the 'Blue Mountain Eye Study' (BMES) study of older Australians, individuals with high omega-3 fat intake had the lowest risk of ARM (OR = 0.41, 0.22 – 0.75) (Chua et al., 2006). There was no association, however, between ARM and butter, margarine, or various nutrients. Abdominal obesity may also be a risk factor for AMD especially in males (Adams et al., 2011). In addition, studies have suggested the ratio of various types of fatty acid in the diet may be important. Hence, AMD patients were divided into five clinical groups and differences in the ratio of omega-6 to omega-3 fatty acids examined. There were significant differences in the ratio in patients at all stages of wet AMD suggesting a decreased ratio may be protective (Mance et al., 2011). In a further study, it was concluded that consumption of omega-3 fatty acids and antioxidants may be an effective preventative strategy (Kishan et al., 2011). AMD has been related to various measures of cardiovascular health. Hence, high mean arterial blood pressure (BP) was associated with an increased risk of AMD (OR = 1.37, 1.04 – 1.79) (Ngai et al., 2011). In the Singapore Malay Eye Study (SiMES), however, no significant association between retinal arteriolar wall signs and early AMD was observed (Cheung et al., 2011). Nevertheless, arteriolar wall opacification was associated with the presence of soft drusen (OR = 1.58) especially in the eyes of non-statin users. By contrast, Klein et al. (2003) found no association between ARM and hypertension. The association between cardiovascular risk factors and AMD has been studied in an Austrian population (Haas et al., 2011). Of 19 cardiovascular risk factors, however, none appeared to be statistically associated with AMD. Nevertheless, the authors concluded that the proteins beta-fibrinogen and apolipoprotein E should be studied further as possible risk factors. A further association with AMD has been observed in patients with chronic kidney disease (Choi et al., 2011), a condition characterized by reduced glomerular filtration, such patients exhibiting an increased risk of early AMD (OR = 1.68, 1.04 – 2.72) and the presence of peripheral retinal drusen (OR = 2.01, 1.02 – 3.99). By contrast, Cackett et al. (2011) studied risk factors for AMD in a Chinese population and found that smoking but not vascular factors were significantly associated with the condition. Similarly, Kabasawa et al. (2011) found in the Japanese population that although smoking increased the risk of AMD, serum fatty acid levels were not related to AMD. Hence, the relationship between AMD, diet, and cardiovascular health remains controversial and there are likely to be many synergistic effects between these various factors.

Sunlight

Whether exposure to sunlight is a significant risk factor for AMD is controversial (Chalam et al., 2011). Although visual perception in humans occurs between wavelengths 380 - 760 nm, the eyes are also sensitive to UV (100 – 380 nm) and IR (770 – 10,000 nm) light. Most UVA and UVB of wavelength less than 295 nm is absorbed and blocked by the cornea and lens, but a fraction at 320 nm reaches the retinal pigment epithelium (RPE) cells. The retina is also vulnerable to visible light, wavelengths between 400 and 760 nm resulting in chronic damage when absorbed by retinal proteins such as melanin, retinal, and lipofuscin (Gaillard et al., 1995; Ragauskaite et al., 2001). The RPE is highly active and prone to oxidative stress and can be damaged especially by UV and blue light. UV and blue light (Margrain et al., 2004) cause photooxidative stress by the generation of 'reactive oxygen species' (ROS) resulting in DNA damage, and the alteration of size and movement of RPE cells (Roduit and Schorderet, 2008).

The retina possesses inherent protection against photochemical damage via possession of antioxidant enymes such as superoxidase dismutase (SOD), catalase, and glutathione peroxidase (Mainster. 1987; Reme et al., 1996; Feenay and Berman, 1976), macular pigments such as melanin, haemoglobin, and flavoproteins which absorb light, and the fact that photoreceptor cells can shed potentially damaged outer segment discs (Marshall, 1976; Young, 1976). Macular pigment ocular density (MPOD) has been measured in healthy subjects using heterochromatic flicker photometry (Yu et al., 2012). MPOD were 0.56, 0.49, 0.36, and 0.19 at eccentricities 0.25, 0.5, 1.0, 1.75 respectively, significantly declining with age, densities being greater in males than females. MPOD levels in healthy elderly subjects may also be positively correlated with anti-oxidant blood markers such as glutathione (Qin et al., 2011).

Epidemiological evidence for light exposure as a cause of AMD is inconclusive (Chalam et al., 2011) and only a few studies report a positive association (Appendix 3). Hence, in the Beaver Dam Eye Study (BDES), the amount of time spent outdoors was positively associated with AMD (OR = 2.26, 1.06 – 4.81) (Cruickshanks et al., 2001). In addition, exposure to summer sun (>5 hours per day in the teenage years) increased retinal pigment (RR = 3.17, 1.24 – 8.11) and early ARM (RR = 2.14, 0.99 – 4.61) (Tomany et al., 2004). In the Chesapeake Bay Waterman Study (CBWS), long-term high exposure to blue light was associated with increased advanced AMD (OR = 1.36, 1.00 – 1.85) (Taylor et al., 1992). Vojnikovic et al. (2010) studied sun exposure and visual field damage in a region of Russia. They concluded that increased sun exposure may represent an important risk factor for AMD and suggested that the eyes of children should be protected by taking vitamin A and antioxidants. Finally, Wielgus et al. (2010) studied the effect of blue light on rat eyes, an experimental model of dry AMD, and found that short wavelength blue light induced retinal injury and should therefore be considered a risk factor for AMD. Several other case-control studies, however, have failed to show associations between sunlight exposure and AMD (West et al., 1989; Wang et al., 2000; Delcourt et al., 2001; Klein, 2007). For example, a study of 446 subjects suggested that sunlight may not be a risk factor for AMD (Khan et al., 2006). In addition, no associations were found between iris or hair colour and early or late ARM (Wang et al., 2003) although very fair skinned individuals had increased risk of developing dry AMD. However, in the BDES study, correcting for age and gender, brown eyes were significantly more likely to develop soft indistinct drusen (RR = 1.53, 1.19 – 1.97) compared with blue eyes (Tomany et al., 2003), but less likely to develop retinal pigment epithelial degeneration (RR = 0.58, 0.41 – 0.82). Hence, many aspects of the relationship between sunlight and AMD remain controversial (Chalam et al., 2011).

Smoking

A summary of the most important epidemiological studies of AMD and smoking is shown in Appendix 4. The earliest case-control studies often report inconsistent results. Hence, a Baltimore study reported a significant relationship between AMD and smoking in males (OR = 2.6) (Hyman et al., 1983), a Maryland study, (West et al., 1989) an increased risk of AMD in smokers, while a hospital-based study carried out in London, UK and a study carried out in Finland (Hirvela et al., 1996) found no relationship with smoking (Pauleijkhoff et al., 1992).

Epidemiological studies have also attempted to relate smoking to the risk of dry and wet AMD. Hence, the Macular Photocoagulation Study Group (MPSG) (1986) found that wet AMD recurred more frequently in treated patients who smoked more than 10 cigarettes a day. Furthermore, the Eye Disease Case-Control Study Group (EDCCSG) (1992) found an increased risk of wet AMD in current smokers (OR = 2.6) (Yannuzzi et al., 1992), while in the BDES, (Klein et al., 1993), no correlation between dry AMD and smoking was observed, although a positive association with ARM was detected in both males and females. Vinding et al. (1992), in a Copenhagen study, found a close relationship between smoking and dry AMD (OR = 2.5) while the BMES, located in Sydney (Smith et al., 1996), found that smoking was strongly related to both wet and dry AMD (OR = 3.92), current smokers having the highest risk.

A number of population-based studies have investigated further relationships between smoking behavior and AMD. In the Pathologies Oculaires Liées à l'Age study (POLA), Delcourt et al., (1998) estimated the risk of AMD in current smokers (OR = 3.6, 1.1 – 12.4) and also demonstrated in ex-smokers, that the risk of AMD may remain up to 20 years after the cessation of smoking (OR = 1.3, 0.4 – 4.3). In a Japanese study, Tamakoshi et al. (1997) found that the risk was greater for current smokers (OR = 2.97, 1.00 – 8.84) compared with ex-smokers (OR = 2.09, 0.71 – 6.13). Bressler et al. (2000) used stereoscopic colour fundus photographs of the macula and divided patients into five separate groups which varied in the severity of AMD. A significant relationship with smoking, however, was found in only one of these groups.

Follow-up studies have been reported on some groups of patients. Hence, further observation of the BDES group (Mitchell et al., 1999) showed that smoking in males and females was associated with the presence of large drusen, a principal pathological sign of late AMD. In addition, in many of the individual epidemiological studies, patient numbers were too low to assess the importance of risk factors, so data were pooled for meta-analysis. Hence, Smith et al., (2001) combined data from the BDES, Rotterdam, and BMES studies and found that AMD was strongly age-related and smoking the only risk factor consistently associated with AMD from all study sites.

Case-control studies have also employed matched pairs of patients to study the effect of smoking on AMD. For example, De Angelis (2004) studied 73 sibling pairs in which one member of each pair exhibited wet AMD in at least one eye while the unaffected sibling was normal and was past the age at which their partner was diagnosed with AMD. A modest 2% increase in the risk of wet AMD was associated with each pack-year of smoking (OR = 1.02, 1.01 – 1.04).

In a prospective study, Seddon et al. (1996) studied female registered nurses 50 – 59 years of age in the USA and found that individuals who smoked more than 25 cigarettes a day had an increased risk (RR = 2.4, 1.4 – 4.0) of developing AMD. The risk did not decline until 15 years after the patient had stopped smoking. In a study of male doctors (Christen et al., 1996), smoking increased the risk of AMD with a 2-3 times increase in men who were current smokers of 20 or more cigarettes a day (RR = 2.46, 1.60 – 3.79). For the wet form of AMD, smoking appeared to double the risk. However, in a Rotterdam study (Vingerling et al., 1996), the risk of wet AMD was 3.2 times for former smokers, and 6.6 times for current smokers. In a prospective study based on a retinal practice (Seddon et al., 2003), 261 individuals greater than 60 years of age were clinically examined for signs of non-advanced AMD, or a visual acuity of 20/200 or better in at least one eye. The average follow-up time in the study was 4.6 years. The RR for smoking was in the range 1.48 to 1.99 but was not statistically significant. Whole body and abdominal obesity, however, were associated with an increased risk of progression to advanced AMD.

Alcohol

In one of the earliest studies of the relationship of alcohol to AMD (BDES, Ritter et al., 1995), consumption of beer was associated with retinal pigment degeneration (OR = 1.03, 1.02 – 1.28) and wet AMD (OR = 1.41, 1.05 – 1.88), but neither wine nor spirits were related to early or late ARM (Appendix 5). By contrast Smith and Mitchell (1996) concluded that there may be an association between the consumption of spirits and AMD. In addition, in a study of smoking and alcohol (Coleman et al., 2010), alcohol consumption was associated with early AMD (OR = 1.57, 1.18 – 2.11) and there was also an increased risk in 80 year old smokers (OR = 5.49, 1.57 – 19.20), suggesting that even older individuals should stop smoking. Several other studies, however, have failed to find significant associations between alcohol and AMD. Hence, in the Physician's Health Study (PHS) (Ajani et al., 1999), there was no appreciable risk of AMD in those individuals consuming one or more alcoholic drinks a week (RR = 0.97, 0.78 – 1.21). Similarly, in the Beijing Eye Study (BES) (Xu et al., 2009) and a prospective Rotterdam study (Boekhoorn et al., 2008), there was no evidence that alcohol was a significant factor influencing the risk of AMD. Nevertheless, there is evidence that heavy alcohol consumption may be a risk. Hence, Wang et al. (2008) found that heavy alcohol consumption was weakly related to AMD while in a prospective BDES study (Knudtson et al., 2007), heavy drinking, i.e., more than four alcoholic drinks per day, was related to dry AMD (OR = 9.2, 1.7 – 51.2) but moderate drinking was unrelated to AMD. Similarly, Fraser-Bell et al. (2006) found that heavy consumption of alcohol in Lantinos was associated with an increase in both wet (OR = 5.8) and dry (OR = 12.7) AMD.

Conclusion

It is anticipated that there will be considerably more individuals who will suffer low vision or blindness by the year 2020. The four major causes of vision loss and blindness are cataract, diabetic retinopathy, glaucoma, and AMD and all four conditions may involve changes in the microcirculation of eye structures. Hence, many approaches to eye health also incorporate attention to individuals' life style choices, especially in relation to smoking, diet, and cardiovascular health and this approach is very relevant to AMD. Of the large number of risk factors investigated, however, few have been consistently associated with AMD, most notably age, smoking, and genetic susceptibility (Evans, 2001; Klein et al., 2004).

What Are the Major Risk Factors for AMD?

The most important genes implicated in AMD appear to be those involved in immune modulation and the complement system such as CFH (Nischler et al., 2011; Edwards et al., 2005; Klein et al., 2005; Haines et al., 2005; Hageman et al., 2005), *CFB* (Gold et al., 2006), *C3* (Matter et al., 2006, Yates et al., 2007), and *SERPING1* (Gibson et al., 2010) strongly suggesting that inflammatory processed are important in AMD. It has been concluded that SNP in *CFH*, *C3*, *HTRA1*, and *SERPING1* together could account for 45% of the risk of developing AMD (Gibson et al., 2010). Genes associated with membrane transport, e.g., *ABCR* (Stroyer et al., 1999; Dela Paz et al., 2000; Souied et al., 2000), and *CACNG* (Spencer et al., 2011), the vascular system, e.g., *FGF2* (Matter et al., 2006), *LOXL1* (Yang et al., 2006; Haines et al., 2005), and *SELP* (Hageman et al., 2005), and with lipid metabolism, e.g., APOE (McKay et al., 2011; Adams et al., 2012; Halliday et al., 2013), and LIPC (Seddon et al., 2010) are also involved. Of these, the vascular genes are of particular interest because of the involvement of the vasculature in neovascularization. In addition, a series of other putative genes of variable or unknown function could be involved in AMD including *ARMS2* (Kanda et al., 2007; Seddon et al., 2007), *ATP synthase* (Nordgaard et al., 2008), *ERCC6* (Bass et al., 2010), and HTRA1 (Yang et al., 2006, Lee et al., 2010, Stroyer et al., 1999) and it is likely that many further genes will be identified. It has been concluded that SNP in *CFH*, *C3*, *HTRA1*, and *SERPING1* together could account for 45% of the risk of developing AMD (Gibson et al., 2010). In addition, Grassman et al. (2012) constructed a 'genetic risk score' (GRS) for AMD based on 13 'at risk' variations of eight individual genes. AMD patients younger than 75 years of age had a significantly higher GRS score compared with those patients older than 75. Hence, a proportion of cases may be explained by genetics especially in individuals younger than 75. Interactions between different genes and between genes and the environment are also likely to play a prominent role in the development of AMD.

A review of research to date suggests that smoking is the single most important modifiable risk factor for AMD. The significance of smoking as a risk factor for AMD has been controversial (Chan, 1998). Earlier epidemiological studies were often inconsistent and criticised either because of a perceived failure to define the stages of AMD or that statistical measures of a possible relationship between a risk factor and the disease had been misinterpreted (Gottsch et al., 1990). Studies have also varied in experimental design and in patient population. Despite these caveats, studies taken collectively suggest a clear link between smoking and AMD, long-term smoking being associated with approximately a doubling of the risk of late-onset AMD. Other questions relating to the relationship between AMD and smoking behavior including whether smoking is related equally to both wet and dry form of the disease, and the relationship between the intensity of smoking and disease are more controversial.

Many recent studies emphasise the multifactorial nature of AMD and have considered the interactions between different risk factors. Hence, Chakravarthy et al. (2011) studied various clinical risk factors in AMD and concluded that the most important were smoking, previous cataract surgery, and a family history of the disease. Montero et al. (2009) studied the factors associated with wet AMD in Spain. They found that the major risk factors were possession of light coloured irides, smoking in males, and female gender. Cardiovascular risk factors were also considered to be important. Butt et al. (2011) examined multiple risk factors for AMD and found that only age was significantly associated with the disease in both men and women. Nevertheless, in women alone, multivitamin use and total cholesterol also had a significant inverse association with AMD, whereas, sun exposure and high density lipoprotein (HDL) cholesterol had a positive association. In an Australian study (McCarty et al., 2001), patients older than 40 years were examined and three factors identified, viz., age (OR = 2.39, 1.17 – 1.29), smoking (>40 years of age) (OR = 2.39, 1.02 – 5.57) and angiotensin converting enzyme inhibitor medications (ACE) (anti-hypertensive drugs) (OR = 3.26, 1.02 – 5.07). In addition, Kawasaki et al. (2008) in a Japanese study, found that increasing age (OR = 2.27, 1.10 – 4.67), and current cigarette smoking (OR = 5.03, 1.00 – 25.47) were the principal factors associated with late AMD. In a Swiss study (Bauer et al., 2008), smoking explained 7% of late AMD cases while Bai et al., (2008) found no gender effect but reported that advancing age and smoking had the strongest associations with AMD. Hence, smoking remains the only easily controllable risk factor for AMD among many environmental and systemic factors that have been examined.

A risk assessment model has been developed for advanced AMD and is available for online use (Klein et al., 2011). Based on data from 2846 AMD patients followed over 9.3 years, Cox proportional hazard analysis was used to assess the significance of demographic, environmental, phenotypic, and genetic covariates. The final model incorporates age, smoking behavior, family history, and genetic variation (especially of *CFH* and *ARMS2*) to predict the risk of AMD.

Modifiable Risk Factors

Several studies have proposed life style changes to reduce the lifetime risk of AMD. Smoking is the most established risk factor and this knowledge may help individuals to take preventative measures, e.g., some studies suggest that stopping smoking could reduce the risk of recurrence, especially of the wet form of the disease. In addition, a high fat intake may be associated with an increased risk of AMD in both women and men. Fat provides about 42% of the food energy in the average UK diet but a diet that derives closer to 20-25% of total food energy from fat is probably healthier. Reducing fat intake to this level means reducing the consumption of red meat and high-fat dairy products such as whole milk, cheese, and butter. Higher red meat intake has been positively associated with early AMD (OR = 1.47, 1.21 – 1.79) whereas consumption of chicken may be negatively associated with AMD (OR = 0.43, 0.20 – 0.91) (Chong et al., 2009). Consumption of omega-3 fatty acids in combination with anti-oxidants has been suggested as a rationale preventative strategy (Kishan et al., 2011) but no fully developed double blind studies have been carried out to date to test this hypothesis. Eating more cold-water fish rather than red meat and adding nuts to the diet have all been suggested to reduce the risk (Desmettre et al., 2004). For example, Chua et al. (2006) suggested that a 40% reduction in ARM could be achieved by changing to a diet with significantly increased fish consumption (Seddon et al., 2001). In addition, there is evidence that supplementing the intake of DHA and eicosapentaemic acid (EPA) and reducing dietary dGI may be protective against AMD in the 'Age Related Eye Disease Study (AREDS) trial (Chiu et al., 2009).

Various nutritional factors may be involved in the pathogenesis of AMD including anti-oxidants and anti-oxidant co-factors, a number of vitamins, zinc, and anti-free radicals such as lutein. zeaxanthin and components of the membrane of photoreceptor cells (Desmettre et al., 2004). Vitamin D and early AMD have been studied in post-menopausal women (Millen at al., 2011), vitamin D being protective against AMD. In particular, a high serum level of vitamin D may protect against early AMD especially in women younger than 75 years of age. A recent review by Evans and Lawrenson (2012) of all randomized controlled trials comparing anti-oxidant vitamins concluded, however, that there was no benefit of any anti-oxidant combination (RR = 0.98, 0.89 – 1.08 any AMD, RR = 1.08, 0.80 – 1.39 advanced AMD). They also concluded that there was accumulating evidence that neither vitamin E nor β-carotene could prevent or delay the onset of AMD. In addition, Delcourt et al. (2010) studied the relationship between nutrition and AMD and found that individuals taking the antioxidants, lutein, zeaxanthin, and omega-3 poly unsaturated fatty acids had a lower risk of AMD. A similar result was reported by Lecerf and Desmettre (2010). By contrast, Chong et al. (2007) using meta-analysis, studied the effect of dietary anti-oxidants in nine prospective cohorts and three random clinical trials. Pooled results suggested that vitamins A, C, and E, zinc, lutein, zeaxanthin, carotene, P-cryptoxanthin, and lycopene had little or no effect on preventing early AMD. However, there was an indication that antioxidant supplements containing zinc could prevent visual loss in individuals with early AMD (Klein, 2007).

There may be an association between the risk of AMD and levels of various antioxidants in blood. Hence, in smokers, nicotine can reduce the levels of antioxidants and the activity of the enzyme retinal phospholipase A and therefore, these factors could contribute to degeneration of the retina during aging (Sastry and Hemontolos, 1998). Plasma levels of α-tocopherol, a form of vitamin E, were studied in a large French cohort (Delcourt et al., 1999), a negative correlation being found with late-onset AMD. In addition, although no relationship was found with plasma retinol, vitamin C, or red blood cell glutathione, the study suggested that vitamin E may provide some protection against AMD (Delcourt et al., 1999). Hence, an important future question is whether supplementing the diet with antioxidants could also protect the retina from the long-term effects of cigarette smoking. Sun protection may also be a useful strategy to reduce the risk of AMD (Glazer-Hockstein et al.,

1990). Hence, in the BDES study, the use of hats and sunglasses at least 50% of the time was associated with a decrease risk of soft indistinct drusen (RR = 0.55, 0.33 – 0.90) (Tomany et al., 2004). Hence, genetic risk factors are now considered to play a more significant role in AMD than previously thought especially in individuals younger than 75 years of age. The most important genes implicated in AMD appear to be those involved in immune modulation and the complement system such as *CFH*, *CFB*, *C3*, and *SERPING1* strongly suggesting that inflammatory processed are important in AMD. Whether the risk of AMD is increased by long-term exposure to sunlight, diet, and cardiovascular factors is more controversial. Smoking appears to be the strongest modifiable risk factor for AMD, the majority of the studies showing a significant association with AMD. Current smokers are, on average, at 2 - 3 times higher risk of AMD than non-smokers and the risk increases with the intensity of smoking. By contrast, only heavy drinking appears to be appreciably linked to AMD. There are also interactions between various risk factors and the risk of AMD that have been little studied to date. Optometrists as front-line informers and educators of ocular health and need to be aware of the risk factors associated with AMD. Cessation of smoking, eye protection under condition of high light intensities, drinking moderate levels of alcohol, and appropriate dietary changes are all modifiable behavioural changes that can be discussed with patients, especially in smokers or in families with a history of the disease, which may reduce the lifetime risk of AMD.

Appendix 1. Genes associated with age-related macular degeneration (AMD)

Gene	Function	Location	Association with AMD
ATP-binding cassette protein (*ABCR*)	Membrane transport	1p22	Wet (+)
Fibulin 5 (*FBLN5*)	Polymerization of elastin	14q32.1	AMD (+)
ATP synthase 6(*MT-ATP6*)	Produces ATP from ADP	MtDNA	AMD(+)
Age-related maculopathy Susceptibility Protein 2(*ARMS2*)	unknown	10q26.13	Wet(+), Dry (?)
Lysyl oxidase-like1 (*LOXL1*)	Connective tissue development	15q24.1	Wet (+)
Serum	Regulates cell	10q25.2-	Bilateral CNV(+)

Gene	Function	Location	Association with AMD
protease(*HTRA1*)	growth and IGF	26.2	
Complement protein (*c3*)	Complement system	19p13.3-p13.2	Wet (+)
Complement factor	Complement	1q32	Bilateral dry (+)
H (*CFH*)	activation	Wet (+)	
Selection P (*SELP*)	Platelet/leucocyte interactions	1q22-25	Dry(+)
Voltage-dependant calcium channel gram 3 (*CACNG3*)	L-type calcium channel protein	16p12	AMD(+)
Fibroblast growth Factor 2 (*FGF2*)	Diverse biological functions	4q26	Dry(+)
Transferrin(*Tf*)	Iron binding protein	3q29	AMD(+) in smokers
Hepatic lipase(*LIPC*)	Converts IDL	15q21 to LDL	Advanced AMD(-)
DNA excision repair protein(*ERCC6*)	Promotes complex formation at repair sites	10q11.23	AMD(+)
Serpin peptidase Inhibitor(*SERPING1*)	Regulates complement	11q12-q13.1	AMD(+)

Appendix 2. Epidemiological studies relating diet and cardiovascular factors to AMD

Factor	Type	OR/RR	CI	Association with AMD	Reference
dGI	PS	1.10	1.00 – 1.20	AMD (+)	Chiu et al. (2007a)
dGI	CC	1.42	1.09 – 1.84	Large drusen (+)	Chiu et al. (2007b)
Red meat	CC	1.47	1.21 – 1.79	AMD(+)	Chong et al. (2009)
Chicken	CC	0.43	0.20 – 0.91	AMD(-)	Chong et al. (2009)
Total fat	PS	1.54	1.17 – 2.01	AMD (+)	Cho et al. (2001)
Linolenic	PS	1.49	1.15 – 1.94	AMD (+)	Cho et al. (2001)
DHA	PS	0.70	0.52 – 0.93	AMD (-)	Cho et al. (2001)

Appendix 2. (Continued)

Factor	Type	OR/RR	CI	Association with AMD	Reference
Omega-3	CC	0.41	0.22 – 0.75	ARM(-)	Chua et al. (2006
BP	CC	1.37	1.04 – 1.79	AMD (+)	Ngai et al. (2011)
Retinal signs	CC	1.58	1.06 – 2.35	Soft drusen (+)	Cheung et al. (2011)
Kidney	CC	1.68	1.04 2.72	Early AMD (+),	Choi et al. (2011)
disease	CC	2.01	1.02 – 3.99	Peripheral drusen (+)	
Anti-hyper. Medication	CC	3.26	1.02 – 5.07	AMD (+)	McCarty et al. (2001)
BP	CC	0.7	None	Cackett et al. (2011)	

BP = Blood pressure, DHA = docosahexaenic acid, dGI = Glycemic index, CC = Case-control study, OR = Odds ratio, RR = relative risk, CI = 95% confidence intervals, PS = Prospective study.

Appendix 3. Epidemiological studies relating the effects of sunlight and related factors to AMD

Factor	Type	OR/RR	CI	Association with AMD	Reference
Time outdoors	CC	2.26	1.06 – 4.81	AMD (+)	Cruickshanks et al. (2001)
Summer sun	PS	3.17	1.24 – 8.11	Retinal pigment (+),	Tomany et al. (2004)
exposure	PS	2.14	0.99 – 4.61	ARM (+)	
Blue light	CC	1.36	1.00 – 1.85	Advanced AMD (+)	Taylor et al. (1992)
Brown eyes	PS	1.53	1.19 – 1.97	Soft drusen (+)	Tomany et al. (2003)
	PS	0.58	0.41 – 0.82	Pigment degen (-)	
HASR	CC	0.73	0.54 – 0.98	Early AMD (-)	Delcourt et al. (2001)

Factor	Type	OR/RR	CI	Association with AMD	Reference
Fair skin	CC	3.5	1.2 – 10.4	None	Wang et al. (2003)
Iris colour	PS	1.53		Soft drusen (+)	Tomany et al. (2003)
Summer sun exposure	PS	3.17	1.24 – 8.11	Retinal pigment (+)	Tomany et al. (2003)
Sun exposure(teenagers)	PS	2.14	0.99 – 4.61	Early ARM (+)	Tomany et al. (2003)

ARM = Age-related maculopathy, CC = Case-control study, HASR = , PS = Prospective study.

Appendix 4. Epidemiological studies relating smoking to AMD

Type	OR/RR	CI	Association	Reference
CC	2.6 (CS)	-	AMD in males (+)	Hyman et al. (1983)
CC	-	-	AMD (-)	West et al. (1989)
CC	-	-	Wet AMD (+)	MPSG
CC	2.6 (CS)	-	Wet AMD (+)	Yannuzzi et al. (1992)
CC	-	-	Wet AMD (+)	Klein et al. (1993)
CC	-	-	AMD (-)	Pauleikhoff et al. (1992)
CC	2.5	-	Dry AMD (+)	Vinding et al. (1992)
CC	-	-	ARM (-)	Hirvela et al. (1996)
CC	3.92 (CS)	-	Wet and dry AMD (+)	Smith et al. (1996)
CC	2.97 (CS)	1.00 – 8.84	Wet AMD (+)	Tamakoshi et al.
	2.09 (ES)	0.71 – 6.13	Wet AMD (+)	(1997)
CC	3.6 (CS)	1.1 – 12.4	AMD (heavy smokers) (+)	Delcourt et al.(1998)
	1.3 (ES)	0.4 – 4.3		
CC	2.39	1.02 – 5.57	AMD (+)	McCarty et al. (2001)

Appendix 4. (Continued)

Type	OR/RR	CI	Association	Reference
CC	4.06	-	Wet and Dry AMD (+)	Kabasawa et al. (2011)
	4.59	-	Wet AMD (+)	
CC	5.03 (CS)	1.0 – 25.47	Late AMD (+)	Kawasaki et al. (2008)
MPCC	1.02	1.01 – 1.04	Wet AMD (+)	De Angelis et al. (2004)
PS	2.4 (CS)	1.40 – 4.0	Wet and dry AMD (+)	Seddon et al. (1996)
	2.0 (ES)			
PS	2.5 (CS	1.60 – 3.79	AMD (+)	Christen et al. (1996)
	1.30 (ES			
	6.6 (CS)		Wet AMD (+)	Vingerling et al. (1996)
	3.2 (ES)			
PS	1.48-1.99	-	AMD (-)	Seddon et al. (2003)

Abbreviations: Current smokers (CS), ex-smokers (ES), MPCC = Matched pairs case-control study, OR = odds ratio, RR = relative risk.

Appendix 5. Epidemiological studies relating alcohol consumption to AMD

Factor	Type	OR/RR	CI	Association	Reference
Beer	CC	1.03	1.02 – 1.28	Retinal pigment (+)	Ritter et al. (1995)
Alcohol	CC	1.41	1.05 – 1.88	Early AMD (+)	Ritter et al. (1995)
1+ drink per week	PS	0.97	0.78 – 1.21	AMD (-)	Anjani et al. (1999)
Alcohol	CC	1.57	1.18 – 2.11	Early AMD (+)	Coleman et al. (2010)
Heavy consumption	PS	9.2	1.7 – 51.2	Dry AMD (+)	Knudson et al. (2007)
Heavy consumption	CC	5.8	-	Wet AMD (+)	Fraser-Bell et al. (2006)
	CC	12.7	-	Dry AMD (+)	

CC = Case-control study, OR = Odds ratio, RR = relative risk, CI = 95% confidence intervals, PS = Prospective study.

References

Adams MKM, Simpson JA, Aung KZ, Makeyeva GA, Giles GG, English DR, Hopper J, Guymer RH, Baird PN, Robman LD. Abdominal obesity and age-related macular degeneration. *Am J Epidemiol* 2011, 173, 1246-1255.

Ajani UA, Christen WG, Manson JE, Glynn RI, Schaumberg D, Buring JE, Hennekens CH. A prospective study of alcohol consumption and the risk of age related macular degeneration. *Anns Epidemiol.* 1999, 9, 172-177.

Allikmets R, Wasserman WW, Hutchinson. An organisation of the ABCR gene: analysis of promotor and splice junction sequences, *Gene* 1998, 215, 111-122.

Baas DC, Despriet DD, Gorgels TGMF, Bergeron-Sawitzke J, Uitterlinden AG, Hofman A, van Duijn CM, Merriam JE, Smith RT, Barile GR, Ten Brink JB, Vingerling JR, Klaver CCW, Allikmets R, Dean M, Bergen AAB. The ERCC6 gene and age-related macular degeneration. *Plos One,* 2010,5, e13786.

Bai ZL, Ren BC, Yang JG, He YA, Chen L, Sun NX. Epidemiological investigation on age related macular degeneration in rural area of Shaanxi Province, China. *Int J Ophthalmol* 2008,1, 77-84.

Bauer P, Barthelmes D, Kurz M, Fleischhauer JC, Sutter FK. The potential effect of population development, smoking and antioxidant supplementation on the future epidemiology of age-related macular degeneration in Switzerland. *Klin Monats Augenheil* 2008, 225, 376-379.

Boekhoorn SS, Vingerling JR, Hofman A, de Jong PTVM. Alcohol consumption and risk of aging macula disorder in a general population – The Rotterdam Study. *Arch Ophthalmol* 2008, 126, 834-839.

Bressler AR, Davis MD, Ferris FL, Klein R, Lindblad AS, Milton RC, Sperduto RD. Risk factors associated with age-related macular degeneration: age related eye disease report 3. *Ophthamology* 2007, 107, 2224-2232.

Brion M, Sanchez-Salorio M, Corton M, de la Fuente M, Pazos B, Othman M, Swaroop A, Abecasis G, Sobrino B, Carracedo A. Genetic association study of age-related macular degeneration in the Spanish population. *Acta Ophthalmol* 2011, 89, 12-22.

Butt AL, Lee ET, Klien R, Russel D, Ogola G, Warn A, Kingsley RM, Yeh JL. Prevalence and risks factors of age-related macular degeneration in Oklahoma Indians the Vision Keepers Study. *Ophthalmology* 2011, 118, 1380-1385.

Cackett P, Yeo I, Cheung CMG, Vithana EN, Wong D, Tay WT, Tai ES, Aung T, Wong TY. Relationship of smoking and cardiovascular risk factors with polypoidal choroidal vasculopathy and age-related macular degeneration in Chinese persons. *Ophthalmology* 2011, 118, 846-852.

Chakravarthy U, Wong TY, Fletcher A, Piault E, Evans C, Zlateva G, Buggage R, Chalam KV, Khetpal V, Rusovici R, Balaiya S. A review: Role of ultraviolet radiation in age-related macular degeneration. *Eye and Contact Lens-Science and Clinical Practice*, 2011, 37, 225-232.

Chalam KV, Khetpal V, Rusovici R, Balaiya S. A review: role of ultraviolet radiation in age-related macular degeneration. *Eye and Contact Lens* 2011, 37, 225-232.

Chan D. Cigarette smoking and age-related macular degeneration. *Optom Vis Sci* 1998, 75, 476-484.

Chen YH, Zeng JX, Zhao C, Wang K, Trood E, Buehler J, Weed M, Kasuga D, Bernstein PS, Hughes G, Fu V, Chin J, Lee C, Crocker M, Bedell M, Salasar F, Yang ZL, Goldbaum M, Ferreyra H, Freeman WR, Kozak I, Zhang K. Assessing susceptibility to age-related macular degeneration with genetic markers and environmental factors. *Arch Ophthamol* 2011, 129, 344-351.

Cheung CMG, Cheoung CYL, Yang EL, Mitchell P, Wang JJ, Wong TY. Retinal arteriolar wall signs and early age-related macular degeneration: the Singapore Malay Eye Study. *Am J Ophthalmol* 2011, 152, 108-113.

Chiu CJ, Milton RC, Klein R, Gensler G, Taylor A. Dietary carbohydrate and the progression of age-related macular degeneration: a prospective study from the age-related eye disease study. *Am J Clin Nutr* 2007a, 86, 1210-1218.

Chiu CJ, Milton RC, Gensler G, Taylor A. Association between dietary glycemic index and age-related macular degeneration in nodiabetic participants in the age-related eye disease study. *Am J Clin Nutr* 2007b, 86, 180-188.

Chiu CJ, Klein R, Milton RC, Gensler G, Taylor A. Does eating particular diets alter the risk of age-related macular degeneration in users of the Age-Related Eye Disease Study Supplements? *Brit J Ophthalmol* 2009, 93, 1241-1246.

Cho E, Hung S, Willett WC, Spiegelman D, Rimm EB, Seddon JM, Colditz GA Hankinson SE. Prospective study of dietary fat and the risk of age-related macular degeneration. *Am J Clin Nut* 2001, 73, 209-218.

Choi J, Moon JW, Shin HJ. Chronic kidney disease, early age-related macular degeneration, and peripheral retinal drusen. *Ophthal Epidemiol* 2011, 18, 259 – 263.

Chong EWT, Wong TY, Kreis AJ, Simpson JA, Guymer RH. Dietary antioxidants and primary prevention of age related macular degeneration: systematic review and meta-analysis. Brit Med J 2007, 335, 755-759.

Chong EWT, Simpson JA, Robman LD, Hodge AM, Aung KZ, English DR, Giles GG, Guymer RH. Red meat and chicken consumption and its association with age-related macular degeneration. *Am J Epidemiol* 2009, 169, 867-876.

Christen WG, Glynn RJ, Mauson JE, Ajain UA, Buring JE. A prospective study of cigarette smoking and the risk of age-related macular degeneration in men. *J Am Med Assoc* 1996, 276, 1147-1151.

Chua B, Flood V, Rochtchina E, Wang JJ, Smith W, Mitchell P. Dietary fatty acids and the 5-year incidence of age-related maculopathy. Arch Ophthalmol 2006, 124, 981-986.

Coleman AL, Seitzman RL, Cummings SR, Yu F, Cauley JA, Ensrud KE, Stone KL, Hochberg MC, Pedula KL, Thomas EL, Mangione CM. The association of smoking and alcohol use with age-related macular degeneration in the oldest old: The study of osteoporotic fractures. Am J Ophthalmol 2010, 149, 160-169.

Cruickshanks KJ, Klein R, Klein BEK, Nondahl DM. Sunlight and the 5-year incidence of early age-related maculopathy: The Beaver Dam eye study. *Arch Ophthamol* 2001, 119, 246-250.

De Angelis MM, Lane AM, Shah CP, Ott J, Dryza TP, Miller JW. Extremely discordant sib-pair study design to determine risk factors for neovascular AMD. *Arch Ophthalmol* 2004, 122, 575-580.

De la Paz MA, Guy VK, Abou-Doria S. Analysis of Stargardt disease gene (ABCR) in age-related macular degeneration. *Ophthalmol* 1999, 106, 1531-1536.

Delcourt C, Diaz JL, Poulon-Sanchez A, Papoz L. Smoking and age-related macular degeneration: The POLA study. *Arch Ophthalmol* 1998, 116, 1031-1035.

Delcourt C, Cristol JP, Tessier F, Leger CL, Descomps B, Papoz L. Age-related macular degeneration and antioxidant status in the POLA study. *Arch Ophthamol* 1999, 117, 1384-1390.

Delcourt C, Carriere I, Ponton-Sanchez A, Fourrey S, Lacroux A, Papez L. Light exposure and the risk of age-related macular degeneration. *Arch Ophthalmol* 2001, 119, 1463-1468.

Delcourt C, Korobelnik JF, Barberger-Gateau P, Delyfer MN, RougiRer MB, Le Goff M, Malet F, Colin J, Dartigues JF. (2010). Nutrition and age-related eye disease: the alienor (antioxydants, lipids essentiels, nutrition et maladies oculaires) study. *J Nut Health & Aging* 2010, 14, 854-861.

Desmettre T, Lecerf JM, Souied EH. Nutrition and age-related macular degeneration. J Fr *Ophthalmol* 2004, 27, 38-56.

Dietzel M, Farwick A, Hense HW. Genetic and risk factors for exudative AMD. *Ophthalmologe* 2010, 107, 1103-1108.

Evans JR. Risk factors for age-related macular degeneration. *Prog Ret Eye Res* 2001, 20, 227-253.

Evans JR, Lawrenson JG. Antioxidant vitamin and mineral supplements for preventing age-related macular degeneration. *Cochrane Database Syst Rev* 2012, 6, CD000253.

Feenay L, Berman ER. Oxygen toxicity: Membrane damage by free radicals. *Invest Ophthalmol Vis Sci* 1976, 15, 789-792.

Fraser-Bell S, Wu J, Klein R, Azen SP, Varma R. Smoking, alcohol intake, estrogen use, and age-related macular degeneration in latinos: The Los Angeles Latino eye study. *Am J Ophthalmol* 2006, 141, 79-87.

Fuse N, Suzuki T, Wada Y. Molecular genetic analysis of ABCR gene in Japanese dry form age-related macular degeneration. *Jap J Ophthalmol* 2000, 44, 245-249.

Fuse N, Mengkegale M, Miyazawa A, Abe T, Nakazawa T, Wakusawa R, Nishida K. polymorphisms in ARMS2 (LOC387715) and LOX1 genes in the Japanese with age-related macular degeneration. *Am J Ophthalmol* 2011, 151, 550-556.

Gaillard ER, Atherton SJ, Eldred G, Dillon, J. Photophysical studies on human retinal lipofuscin. *Photochem Photobiol* 1995, 61, 448-453.

Gerber S, Rozet JM, van de Pol TJR, Hoyng CB, Munnich A, Blankenagel A, Kaplan J, Cremers FPM. Complete exon-intron structure of the retina specific ATP binding transporter gene (ABCR) allows identification of novel mutations underlying Stargardt's disease. *Genomics* 1998, 48, 139-142.

Gibson J, Cree A, Collins A, Lotery A, Ennis S. Determination of a gene and environment risk model for age-related macular degeneration. *Brit J Ophthalmol* 2010, 94, 1382-1387.

Glazer-Hockstein, C; Dunaief JL. Could blue light -blocking lenses decrease the risk of age-related macular degeneration? *Retina* 2006 26, 1-4.

Gottsch JD, Pou S, Bynoe LA, Rosen GM. Hematogenous photosensitisation: a mechanism for the development of age-related macular degeneration. *Inv Ophthal Vis Sci* 1990, 31, 1674-1682.

Grassman F, Fritsche LG, Keilhauer CN, Heid IM, Weber BHF. Modeling the genetic risk in age related macular degeneration. Plos One 2012, 7, e37979.

Haas P, Aggermann T, Stiendl K, Krugluger W, Puhringer H, Oberkanins C, Frantal S, Binder S. Genetic cardiovascular risk factors and age related macular degeneration. *Acta Ophthalmol* 2011, 89,335-338.

Hirvela H, Luukinen H, Laara E, Laatikainen L. Risk factors of age-related maculopathy in a population 70 years of age or older. *Ophthalmol (Rochester),* 1996, 103, 871-877.

Hyman LG, Lilienfeld AM, Ferris FL, Fine SI. Senile macular degeneration: a case-control study. *Am J Epidemiol* 1983, 118, 213-227.

Kabasawa S, Mori K, Horie-Inoue K, Gehlbach PL, Inoue S, Awata T, Katayama S, Yoneya S. Association of cigarette smoking but not serum fatty acids with age-related macular degeneration in Japanese population. *Ophthalmology*, 2011, 118, 1082-1088.

Kawasaki R, Wang JJ, Ji GJ, Taylor B, Oizumi T, Daimon M, Kato T, Kawata S, Kayama T, Tano Y, Mitchell P, Yamashita H, Wong TY. Prevalence and risk factors for age-related macular degeneration in an adult Japanese population. *Ophthalmology* 2008, 115, 1376-1381.

Khan, JC; Shahid H, Thurlby DA, Bradley M, Clayton DG, Moore AT, Bird AC, Yates JR, Genetic Factors in AMD Study. Age related macular degeneration and sun exposure, iris colour, and skin sensitivity to sunlight. *Brit J Ophthalmol* 2006, 90, 29-32.

Kishan AU, Modjtahedi BS, Martins EN, Modjtahedi SP, Morse LS. Lipids and age-related macular degeneration. *Surv Ophthalmol* 2011, 56, 195-213.

Klein R. Overview of progress in the epidemiology of age-related macular degeneration. *Ophthalmic Epidemiol* 2007, 14, 184-187.

Klein R, Klein BE, Linton KL, DeMets DL. The Beaver Dam Eye Study: the relation of age-related maculopathy to smoking. *Acta Epidemiol* 1993, 137, 190-200.

Klein R, Klein BEK, Marino EK, Kuller LH, Furberg C, Burke GL, Hubbard LD. Early age-related maculopathy in the cardiovascular health study. *Ophthalmology* 2003, 110, 25-33.

Klein R, Peto T, Bird A, Vannewirk MR. The epidemiology of AMD. *Am J Ophthalmol,* 2004, 137, 486-495.

Klein ML, Francis PJ, Ferris FL, Hamon SC, Clemons TE. Risk assessment model for development of advanced age-related macular degeneration. *Arch Ophthalmol* 2011, 129, 1543-1550.

Knudtson MD, Klein R, Klein BEK. Alcohol consumption and the 15-year cumulative incidence of age-related macular degeneration. *Am J Ophthalmol* 2007, 143,1026-1029.

Lecerf J, Desmettre T. Nutrition and age-related macular degeneration. *J Franc D'Ophthalmol* 2010, 33, 749-757.

Lee SJ, Kim NR, Chin HS. LOC387715/HTRA1 polymorphisms, smoking and combined effects on exudative age-related macular degeneration in a Korean population. *Clin Exp Ophthalmol* 2010, 38, 698-704.

Liu MM, Agron E, Chew E, Meyeole C, Ferris FL, Chan CC, Tuo JS. Copy number variations in candidate genes in neovascular age-related macular degeneration. *Invest Ophthalmol Vis Sci* 2011, 52, 3129-3135.

Maguire MG. Recurrent choroidal neovascularization after argon laser photocoagulation for neovascular maculopathy. *Arch Ophthamol* 1986, 104, 503-512.

Mainster MA. Light and macular degeneration: A biophysical and clinical prospective. *Eye* 1977, 1, 304-310.

Maller JB, Fagerness JA, Reynold RC, Neale BM, Daly MJ, Seddon JM. Variation in complement Factor 3 is Associated with Risk of Age Related Macular Degeneration. *Nature Genetics* 2007, 39, 1200-1201.

Mance TC, Kovacevic D, Alpeza-Dunato Z, Stroligo MN, Brumini G. The role of omega-6 to omega-3 ratio in development and progression of age-related macular degeneration. *Coll Antropol* 2011, 35, 307-310.

Margrain, TH; Boulton M, Marshall J, Sliney DH. Do blue light filters confer protection against age-related macular degeneration*? Prog Ret Eye Res* 2004, 23, 523-31.

Marshal J. Radiation and the aging eye. *Ophthal Physiol Opt* 1976, 5, 241-263.

McCarty CA, Mukesh BN, Fu CL, Mitchell P, Wang JJ, Taylor HR. Risk factors for age-related maculopathy: The visual impairment project. *Arch Ophthalmol* 2001, 119, 1455-1462.

Millen AE, Voland R, Sondel SA, Parekh N, Horst RL, Wallace RB, Hageman GS, Chappell R, Blodi BA, Klien ML, Gehrs KM, Sarto GE, Mares JA. Vitamin D status and early age-related macular degeneration in postmenopausal women. *Arch Ophthalmol* 2011, 129, 481-489.

Mitchell P, Chapman S, Smith W. Editorial: Smoking as a major cause of blindness: A new cigarette pack warning. *MJA* 1999, 171, 173-174.

Montreo JA, Sanchez-Tocino H, Maldonado MJM, Bartolom RMP, Coomonte AV. Epidemiological factors associated with exudative age-related macular degeneration in Spain. *Inter J Ophthalmol* 2009, 2, 74-76.

Mullins RF, Skele JM, Folk JC, Solivan-Timpe FM, Oetting TA, Huang J, Wang K, Stone EM, Fingert JH. Evaluation of variants in the selectin genes in age-related macular degeneration. *BMC Med Genet* 2011, 12, Article 56.

Ngai LY, Stocks N, Sparrow JM, Patel R, Rumley A, Lowe G, Smith GD, Ben-Shlomo Y. The prevalence and analysis of risk factors for age-related macular degeneration: 18 year follow-up data from the Speedwell eye study, United Kingdom. *Eye* 2011, 25, 784-793.

Nischler C, Oberkofter H, Ortnev C, Paikl D, Riha W, Lang N, Patsch W, Egger SF. Complement factor H Y402H gene polymorphism and response to intravitreal bevacizumab in exudative age related macular degeneration. *Acta Ophthalmol* 2011, 89, 344-349.

Nordgaard CL, Karunadharma PP, Feng X, Olsen TW, Ferrington DA. Mitochondrial proteomics of the retinal pigment peithelium at progressive stages of age-related macular degeneration. *Inv Ophthalmol Vis Sci* 2008, 49, 2848-2855.

Nasonkin I, Illing M, Koehler MR, Schmid M, molday RS, Weber BHF. Mapping of the rod photoreceptor ABC transporter (ABCR) to 1p21-1p22.1 and identifications of novel mutations in Stargardt's disease. *Hum Genet* 1998, 102, 21-26.

Owen CG, Jarrar Z, Wormald R, Cook DG, Fletcher AE, Rudnicka AR. The estimated prevalence and incidence of late stage age related macular degeneration in the UK. *Brit J Ophthalmol* 2012, 96, 752-756.

Papaioannou M, Ocaka L, Bessant D. Analysis of ABCR mutations in British patients with recessive retinal dystrophies. *Inv opthal Vis Sci* 2000, 41, 16-19.

Pauleikhoff D, Wormald RP, Wright L, Wessing A, Bird AC. Macular disease in an elderly population. *Ger J Ophthalmol* 1992, 1, 12-15.

Qin L, Bartlett H, Griffiths HR, Eperjesi F, Armstrong RA, Ghergel D. Macular pigment optical density is elated to blood glutathione levels in healthy individuals. *Invest Ophthalmol Vis Sci* 2011, 52, 5029-5033.

Ragauskaite L, Heckathorn R, Gaillard ER. Environmental effects on the photochemistry of A2-E, a component of human retinal lipofuscin. *Phytochem photobiol* 2001, 74, 483-488.

Reme C, Reinboth J, Clausen M, Hafezi F. Light damage revisited: Converging evidence, diverging views? *Graefes Arch Clin Exp Ophthalmol* 1996, 234, 2-11.

Ritter LL, Klein R, Klein BEK, Maresperlman JA, Jensen SC. Alcohol-use and age-related maculopathy in the Beaver Dam Eye Study. *Am J Ophthamol* 1995, 120, 190-196.

Roduit R, Schorderet DF. MAP kinase pathways in UV-induced apoptosis of retinal pigment epithelium ARPE-19 cells. *Apoptosis* 2008, 13, 343-353.

Sastry BVR, Hemontolos ME. Influence of nicotine and cotinine on retinal phospholipase A (2) and its significance to macular function. *J Occ Pharm Therapeut* 1998, 14, 447-458.

Seddon JM, Willett WC, Speizer FE, Haukinson SE. A prospective study of cigarette smoking and age-related macular degeneration in women. *J Am Med Assoc* 1996, 276, 1141-1146.

Seddon JM, Rosner B, Sperduto RD, Yannuzzi L, Haller JA, Blair NP, Willett W. Dietary fat and the risk for advanced age-related macular degeneration. *Arch Ophthamol* 2001, 119, 1191-1199.

Seddon JM, Cote J, Davis N, Rosner B. Progress of age-related macular degeneration: association with body mass index, waist circumference and waist-hip ratio. *Arch Ophthalmol,* 2003, 121,785-792.

Seddon JM, Gensler G, Milton RC, Klein ML, Rifai N. Association between C-reactive protein and AMD. *J Am Med Assoc* 2004, 291, 704-710.

Seddon JM, Reynolds R, Rosner B. Associations of smoking, body mass index, dietary lutein, and the LIPC genetic variant rs10468017 with advanced age-related macular degeneration. *Molecular Vision,* 2010, 16, 2412-2424.

Shroyer NF, Lewis RA, Allikmets R. The rod photoreceptor ATP –binding cassette transport gene (ABCR) and retinal disease: from monogenic to multifactorial. *Vis Res* 1999, 39, 2537-2544.

Smith W, Mitchell P. Alcohol intake and age-related maculopathy. *Am J Ophthalmol* 1996, 122, 743-745.

Smith W, Assink J, Klein R, Mitchell P, Klaver CCW, Klein BEK, Hofman A, Jensen S, Wang JJ, De Jong PTVM. Risk factors for AMD: Pooled findings from three continents. *Ophthalmol* 2001, 108, 697-704.

Smith W, Mitchell P, Leeder S. Smoking and age-related maculopathy. The Blue Mountains Eye Study. *Arch Ophthamol* 1996, 114, 1518-1523.

Souied EH, Ducroq D, Rozet JM. ABCR gene analysis in familial exudative age-related macular degeneration. *Inv Ophthalmol Vis Sci* 2000, 41, 244-247.

Spencer KL, Olson LM, Schnetz-Boutaud N, Gallins P, Wang GF, Scott WK, Agarwal A, Jakobsdottir J, Conley Y, Weeks DE, Gorin MB, Pericak-Vance MA, Haines JL. Dissection of chromosome 16p12 linkage peak suggest a possible tole for CACNG3 variants in age-rleated macular degeneration susceptibility. *Inv Ophthamol Vis Sci* 2011, 52, 1748-1754.

Stein JD, Van der Beek BL, Talwar N, Nan B, Musch DC, Zacks DN. Rates of nonexudative and exudative age-related macular degeneration among Asian American Ethnic groups. *Invest Ophthalmol Vis Sci* 2011, 52, 6842-6848.

Stine GJ. The New Human Genetics, Wm c Brown, Dubuque, Iowa, 1989.

Stone EM, Braun TA, Russel SR, Kuehn MH, Lottery AJ, Moore PA, Eastman CG, Casavant TL, Sheffield VC. Missense variations in the fibulin 5 gene and age-related macular degeneration. *New Eng J Med* 2004, 351, 346-353.

Sun H, Nathans J. Stargardt's ABCR is localized to the disc membrane of retinal rod outer segments. *Nat Genet* 1997, 17, 15-16.

Tamakoshi A, Yuzawa M, Matsui M, Uyama M, Fujiwara NK, Ohno Y. Smoking and the neovascular form of age-related macular degeneration in late middle aged males: findings from a case-control study in Japan. *Brit J Ophthalmol* 1997, 81, 901-904.

Taylor HR, West S, Munoz B, Rosenthal FS, Bressler SB, Bressler NM. The long-term effects of visible light on the eye. *Arch Ophthalmol* 1992, 110, 99-104.

Tomany SC, Klein R, Klein BEK. The relationship between iris color, hair color, and skin sun sensitivity and the 10-year incidence of age-related maculopathy: The Beaver Dam Eye Study. *Ophthalmology* 2003, 110, 1526-1533.

Tomany SC, Cruickshanks KJ, Klein R, Klein BEK, Knudtson MD. Sunlight and the 10-year incidence of age-rlated maculopathy: The Beaver Dam eye study. *Arch Ophthalmol* 2004, 122, 750-757.

Tong Y, Liao J, Zhang YA, Zhou J, Zhang HY, Mao M. LOC387715/HTRA1 gene polymorphisms and susceotibility to age-related macular degeneration: A HuGE review and meta-analysis. *Molecular Vision* 2010, 16, 1958-1981.

Van de Beek BL, Zacks DN, Talwar N, Nan B, Musch DC, Stein JD. Racial differences in age-related macular degeneration rates in the Uinted States: a longitudinal analysis of a managed care network. *Am J Ophthalmol* 2011, 152, 273-282.

Vinding T, Appleyard M, Nyboe J, Jensen G. Risk factor analysis for atrophic and exudative age-related macular degeneration. An epidemiological study of 1000 aged individuals. *Acta Ophthalmol* 1992, 70, 66-72.

Vingerling JR, Hofman A, Groblee DE, De Jong PT. Age-related macular degeneration and smoking: The Rotterdam study. *Arch Ophthalmol,* 1996, 114, 1193-1196.

Vojnikovic B, Synek S, Micovic V, Telezar M, Linsak Z. Epidemiological study of sun exposure and visual field damage in children in Primorsko-Goranska county- the risk factors of earlier development of macular degeneration. *Collegium Antropologicum* 2010, 34, 57-59.

Wang JJ, Foran S, Mitchell P. Age-specific prevalence and causes of bilateral and unilateral visual impairment in older Australians: The Blue Mountains Eye Study. *Clin Exp Ophthalmol* 2000, 107, 875-879.

Wang JJ, Jakobsen K, Smith W, Mitchell P. Five-year incidence of age-related maculopathy in relation to iris, skin or hair colour, and skin sensitivity: The Blue Mountains Eye Study. *Clin Exp Ophthalmol* 2003, 31, 317-321.

Wang SQ, Wang JJ, Wong TY. Alcohol and eye diseases. *Surv Ophthalmol* 2008, 53, 512-525.

West SK, Rosenthal FS, Bressler NM, Bressler SB, Munoz B, Fine SL, Taylor HR. Exposure to sunlight and other risk factors for age-related macular degeneration. *Arch Ophthalmol* 1989, 107, 875-879.

Wielgus AR, Collier RJ, Martin E, Lih FB, Tomer KB, Chignel CF, Roberts JE. Blue light induced A2E oxidation in rat eyes- experimental animal model of dry AMD. *Photochem Photobiol Sci* 2010, 9, 1505-1512.

Winsper SJ, Armstrong RA, Hodgkins PS, Blair JA. Gallium-transferrin binding in treated and untreated Parkinson's disease. *NeuroReport* 1997, 8, 709-711.

Wysokinski D, Szaflik J, Sklodowska A, Kolodziejska U, Dorecka M, Romaniuk D, Wozniak K, Blasiak J, Szaflik JP. The A allele of the-576G>A polymorphism of the transferrin gene is associated with increased risk of age-related macular degeneration in smokers. *Tohoku J Exp Med* 2011, 223, 253-261.

Xu L, You QS, Jonas JB. Prevalence of alcohol consumption and risk of ocular diseases in a general population: The Beijing Eye Study. *Ophthalmol* 2009, 116, 1872-1879.

Yang Z, Camp NJ, Sun H, Tong Z, Gibbs D, Cameron DJ, Chen H, Zhao Y, Pearson E. A variant of the HTRA1 gene increases susceptibility to age-related macular degeneration. *Science*, 2006, 314, 992-993.

Yannuzzi LA, Sorenson JA, Sobel RS, Daly JR, Derosa JT, Seddon JM, Gragoudas ES, Puliafito CA, Gelles E, Gonet R, Burton TC, Culver J, Metzger K, Kalbfleisch N, Zarling D, Farber MD, Blair N, Stelmack T, axelrod A, Waitr SE, Cross A, Rolnick C, Flom T, Haller J, Pusin S, Cassel G, Applegate CA, Seigel D, Sperduto RD, Hiller R, Mowery R, Chew E, Tamboli A, Miller DT. Sowell AL, Gunter EW, Dunn M, Seddon JM, Shamban K, Gelles E, Lento D, Alexander JA, Phillips DA. Risk factors for neovascular age-related macular degeneration. *Arch Ophthalmol* 1992, 110, 1701-1708.

Yates JR, Sepp T, Matharu BK, Khan JC, Thurlby DA, Shahid H, Clayton DJ, Hayward C, Morgan J, Wright AF, Armbrecht AM, Dhillon B, Deary IJ, Redmond E, Bird AC, Moore AT. Complement C3 Variant and the Risk of Age-Related Macular Degeneration. *Engl J Med* 2007, 357, 553-561.

Yoshida A, Yoshida M, Yoshida S, Shiose S, Hiroishi G, Ishibashi T. Familial cases with age-related macular degeneration. *Jap J Ophthalmol* 2000, 44, 290-295.

Young RW. Visual cells and the concept of renewal. *Invest Ophthalmol Vis Sci* 1976, 15, 700-715.

Yu J, Johnson EJ, Shang F, Lim A, Zhou HY, Cui L, Xu J, Snellingen T, Liu XP, Wang NL, Liu NP. Measurement of macular pigment optical density in a healthy Chinese population sample. *Invest Ophthamol Vis Sci* 2012, 53, 2106-2111.

Zhang K, Kniazeva M, Hutchinson A. The ABCR gene in recessive and dominant Stargardt's disease: a genetic pathway in macular degeneration. *Genomics* 1999, 60, 234-237.

In: Age-Related Macular Degeneration
Editor: Melanie Carter
ISBN: 978-1-63483-329-5

Chapter II

HTRA1 Overexpression Induces the Exudative Form of Age-related Macular Degeneration

Daisuke Iejima, Mao Nakayama and Takeshi Iwata
Division of Molecular and Cellular Biology, National Institute of Sensory Organs, National Hospital Organization Tokyo Medical Center, Tokyo, Japan

Abstract

Age-related macular degeneration (AMD) is a leading cause of vision loss and blindness in the elderly. The dry form is more common and accounts for about 85~90% of AMD patients in US, while Japanese AMD patients predominantly progress to wet-form or polypoidal choroidal vasculopathy (PCV). Recent studies have shown *HTRA1*, a serine protease gene, as major risk factor for wet form AMD (De Wan et al., Science 2006). Furthermore, we reported that the Japanese typical wet form AMD patients showed significant association with *ARMS2/HTRA1* (Goto, Akahori et al., JOBDI 2009).

The purpose of this study is to elucidate the function of *ARMS2/HTRA1* gene promoter in wet-form AMD patients. The promoter sequence experiment showed that a great number of AMD patients had specific indel mutation in 3.8 kb upstream of *HTRA1* gene. 2~3-fold increase of promoter activity was observed in indel *HTRA1* promoter compared to control sequence (Iejima et al., JBC 2015). Furthermore, we created transgenic mice ubiquitously overexpressing mouse *HtrA1* using the chicken act in promoter, continuous induction of *HtrA1 in vivo* was shown to lead to CNV, similar to wet AMD patients (Nakayama, Iejima et al., IOVS 2014).

These results suggest that human *HTRA1* expression is enhanced by AMD specific indel mutation in the promoter region of *HTRA1* gene, and this enhanced *HTRA1* may be concerned with induce retinal neovasucularization.

1. Introduction

Age-related macular degeneration (AMD) is the most prevalent cause of legal blindness among elderly in the developed world [1] and among those over 65 years of age [2]. This disease is triggered by a complex mechanism involving genetic and environmental factors (e.g., aging, iris color, smoking, dietary fat, hypertension, and alcohol intake) [3-7]. AMD is classified as either "dry" or "wet" according to the phenotypic characteristics of geographic atrophy and/or choroidal neovascularization (CNV). Frequencies of dry and wet AMD cases vary among ethnic groups [8, 9]. For example, Caucasian patients are predominantly affected by dry AMD, whereas Japanese and other Asian patients are predominantly affected by either wet AMD, polypoidal choroidal vasculopathy (PCV) [9], or retinal angiomatous proliferation (RAP). Among 289 Japanese patients examined with wet AMD, 35.3%, 54.7%, and 4.5% were diagnosed with typical AMD, PCV, and RAP, respectively.

Recent genome-wide association studies (GWASs) have identified more than 19 susceptibility genes associated with AMD [10], including the complement factor H (*CFH*) gene on chromosome 1q32 and the AMD susceptibility 2/High temperature requirement A1 (*ARMS2/HTRA1*)gene on chromosome 10q26. Although the Y402H single nucleotide polymorphism (SNP) in *CFH* (rs1061170) has been strongly associated with dry AMD in Caucasians [11, 12], this locus has not been associated with wet AMD in most Japanese patients [13]. Rather, *ARMS2/HTRA1* is the region is most strongly associated with wet AMD in the latter population [9, 14]. The SNP rs10490924 in a single linkage disequilibrium (LD) block was shown to increase the risk of wet AMD in both Caucasian and Japanese populations [9].

The biological function of *ARMS2*, an evolutionarily recent gene within the primate lineage [15, 16] remains unclear [15-17]. However, it is known that the ARMS2 protein localizes to the mitochondria of the inner segment of the photoreceptor [15] and interacts with fibrin-6 in the extracellular matrix (ECM) of the choroid [17]. *HTRA1* encodes a protein that belongs to the HTRA serine protease family, which is highly conserved from micro- to multicellular organisms, including humans [18]. The HTRA1 protein regulates various cellular processes by cleaving or binding critical factors that participate in cell proliferation, migration, and fate [19-22]. Furthermore, a mutation in *HTRA1* has been associated with a hereditary form of familial cerebral small-vessel disease (CARASIL) [23].

DeWan et al. first described the upregulation of *HTRA1* in the fibroblasts of AMD patients [24]. Subsequently, several groups published confirmatory or conflicting data, and extended the results to the development of a simple promoter characterization assay containing the insertion/deletion (indel) sequence of *HTRA1* [15, 25]. Conclusions from these studies were made on the basis of promoter activity measured in the human retinal pigment epithelium (RPE) cell line ARPE19. More recently, upregulation of human *HTRA1* gene expression in the mouse RPE was shown to induce a branching network of choroidal vessels, polypoidal lesions, and severe degeneration of the elastic lamina and tunica media of choroidal vessels, as seen in AMD and PCV patients [26, 27]. Herein, we investigate whether the pathogenesis of wet AMD is enhanced by *HTRA1* expression derived from a promoter with a mutation in *ARMS2/HTRA1*.

2. Functional Analysis of the *ARMS2/HTRA1* Gene

Previously, we performed a GWAS to analyze the *ARMS2/HTRA1* gene mechanism in Japanese patients with wet AMD, and identified a strong LD across the *ARMS2/HTRA1* region in 10q26 [9]. This LD made genetic association studies alone insufficient to distinguish between cataract patients (as the control group) and patients with wet AMD [9]. Thus, we undertook a comprehensive characterization of wet AMD-associated variants in the region of high LD, and performed a sophisticated analysis of the possible functional relevance of these genes in the disease process.

To determine the genomic sequence of the LD block represented by SNP rs10490924 (*ARMS2* A69S mutation) in association with AMD [9, 24, 28], we sequenced approximately 10.5kbp of the LD block (NC_000010.10) in 228 cataract patients (controls) and 226 AMD patients (Figure 1A). This analysis revealed two unique sequences in the *ARMS2* and *HTRA1* regulatory regions (Figure1B, C) [29]. One sequence included a C/A variant in the promoter region of *ARMS2* located -550bp upstream of exon 1 (Figure 1A, B, D, F). The other sequence was an indel variant in the promoter region of *HTRA1*, located -3,777 bp upstream of exon 1, immediately following *ARMS2* exon 2 (Figure 1A, C, E, G). Sequence variations in both regulatory elements were in complete LD with rs10490924, which suggests that the pathogeneses of AMD and PCV are directly associated with these variants. The indel variant in close proximity to *ARMS2* and *HTRA1* has been at the center of a discussion regarding whether AMD involves one or both genes [24, 28].

Two hypotheses for increased risk of AMD have been proposed to date: increased transcription of *HTRA1* and/or unstable *ARMS2* mRNA [15, 24, 28]. We measured *ARMS2* promoter activity between -1,000 and + 1 base pairs (bp) and between -600 and + 1bp, which both contain the C/A variant (Figure 2A, B, C). We did not detect promoter activity in either the non-C/A or C/A variants of differing lengths in all cell lines tested (Figure 2C).

Next, to test whether the transcription levels of *HTRA1* were altered with expression driven by the indel promoter, we measured the *HTRA1* promoter activity in the -4,320 to + 1bp region of the non-indel promoter and in the -3,936 to + 1bp region of the indel promoter, with a common sequence on both ends of the promoter (Figure 2A, D, E). The indel variant upregulated the promoter activities in the 661W and RGC5 cell lines by 2- and 3-fold, respectively, compared to non-indel promoter activity (Figure 2E). In these cell lines, the *HTRA1* non-indel promoter activity was at undetectably low levels, equivalent to the level of the empty vector. The indel variant did not affect promoter activity in the ARPE19 cell line.

To determine the endogenous level of *HTRA1* expression in individuals with a non-indel promoter sequence, we generated induced pluripotent stem cells (iPSCs) from genotyped normal and AMD individuals who were hetero- or homozygous for the indel promoter sequence. These individuals exhibited a 6- to 7-fold increase in the HTRA1 mRNA transcript compared to individuals without the indel promoter sequence. HTRA1 mRNA expression was not influenced by the heter- or homozygous state of the indel promoter sequence (Figure 3). These results suggest that the indel variant in complete association with rs10490924 is the only major sequence change in this LD block that is likely to play a major role in AMD onset. Furthermore, the *ARMS2* promoter appeared to be only marginal active in all cell lines tested, whereas the hetero- and homozygous forms of the *HTRA1* promoter were robustly upregulated in AMD patient-derived iPSCs. So, these results suggest that human *HTRA1* expression is enhanced by AMD specific indel mutation in the promoter region of *HTRA1* gene.

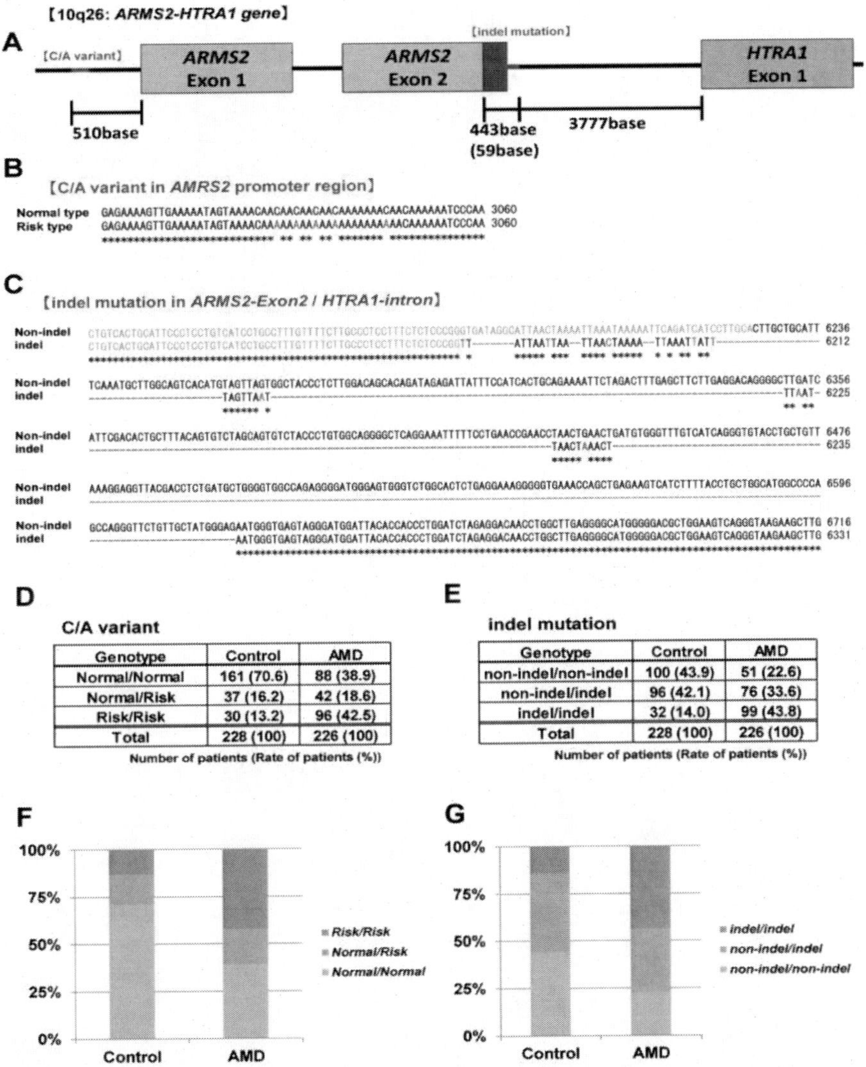

Figure 1. *ARMS2-HTRA1* region in 10q26. Gene structure of *ARMS2* and *HTRA1* (A). Sequences of the C/A variant in the *ARMS2* promoter region (B) and the *ARMS2-HTRA1* insertion/deletion (indel) region (C). Frequencies of the C/A polymorphism (D) and the indel polymorphism (E) in controls and patients with wet AMD. Associations of the homozygous polymorphism in the *ARMS2* promoter region (F) and the homozygous indel polymorphism (G) with risk of wet AMD.

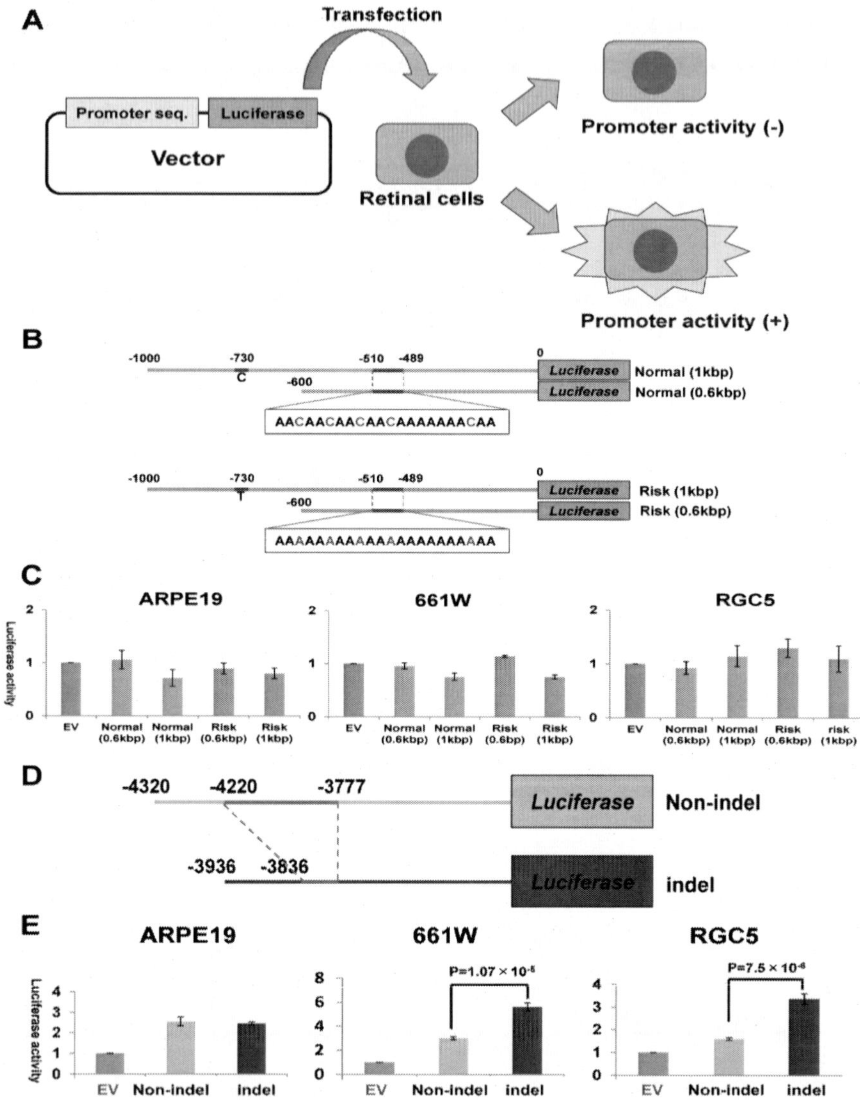

Figure 2. Luciferase assay analysis of the *ARMS2* and *HTRA1* promoter activities. (A) Scheme of the luciferase assay. (B) Four types of luciferase assay vectors were constructed for analysis of the *ARMS2* promoter activity: Normal, 1kbp; Normal, -0.6kbp; Risk, 1kbp; and Risk, -0.6kbp. The 1-kbp sequence contains a C/T SNP at the upper -730bp from *ARMS2* exon1. (C) *ARMS2* promoter activity in retinal cell lines (ARPE19, 661W, RGC5). (D) Two types of luciferase assay vectors were constructed for analysis of the *HTRA1* promoter activity: Non-indel, -4320bp; indel, -3936bp. EV, empty vector control.

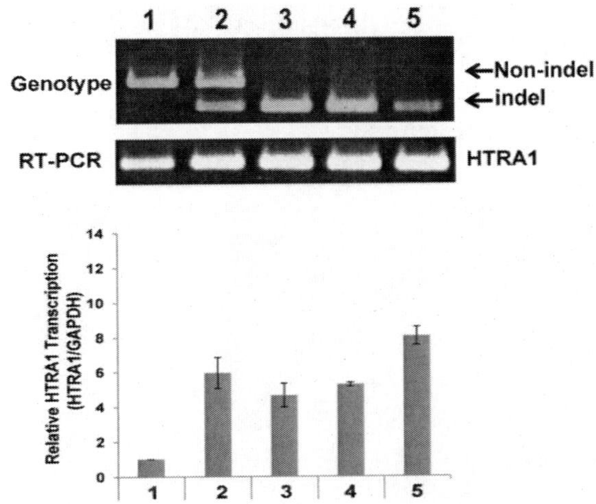

Figure 3. *HTRA1* transcription in human iPSCs derived from patients with wet AMD. Quantitative RT-PCR demonstrates the relative level of *HTRA1* transcription in iPSCs derived from non-indel- and indel-promoter individuals, including a 30-year-old male without AMD (1), 72-year-old female with wet AMD (2), 38-year-old male without AMD (3), 71-year-old female without AMD (4), and 71-year-old female with wet AMD (5).

3. Choroidal Neovascularization Mouse Model

To determine the functions of *ARMS2* and *HTRA1* in the choroid and retina, we constructed human *ARMS2* transgenic (Tg) and mouse *HtrA1* Tg mice (human gene symbol: *HTRA1*, mouse gene symbol: *HtrA1*) (Figure 4, 5) [30]. We observed fluoresce in diapedesis in *HtrA1* Tg mouse fundi, which suggests breakdown of the blood retinal barrier and aggravation of retinal vascular permeability (Figure 4). Lesions with low fluorescence were observed in *HtrA1* Tg mouse choroids by indocyanine green angiography. These lesions indicated the presence of networks of abnormally branched vessels (Figure 4A, left panel). Radial capillary branching from the choroid through the RPE and into the retina was observed by optical coherence tomography (Figure 4A, right panel).

Radial CNV formation was further explored by staining with hematoxylin and eosin (H & E) and an anti-CD31 antibody, which marks endothelial cells (Figure 4B, left and center panel). *HtrA1* Tg mice showed ruptures and deficiencies in Bruch's membranes by elastic-van Gieson staining (Figure 4B, right panel). Four of the 22 twelve-month-old *HtrA1* Tg mice (18.2%), but none of the 40 twelve-month-old wild-type (WT) mice, developed CNV. None of the 20 twelve-month-old *ARMS2* and *ARMS2 (A69S)* Tg mice had any abnormal retinal changes (Figure 5).

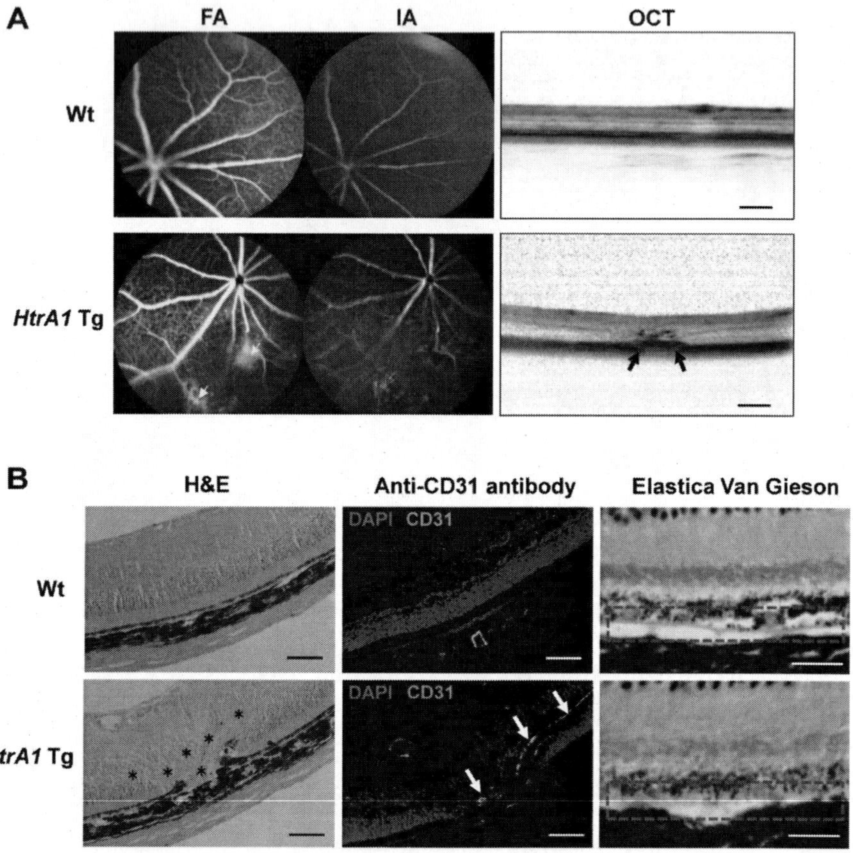

Figure 4. Development of CNV by 12-month-old *HtrA1* Tg mice. (A) Left panel: *HtrA1* Tg mice showed hyperfluorescent lesions by fluoresce in angiography (FA, yellow asterisks) and networks of abnormally branching vessels by indocyanine green angiography (IA, red asterisks). Right panel: Optical coherence tomography (OCT) revealed CNV in the choroid and retina (black arrows; scale bar: 200μm). (B) Left panel: Hematoxylin and eosin (H&E) staining of the FA and IA lesions on the retinal sections showed radial CNV spreading from the choroid through the RPE into the retina (black asterisks; 40× magnification; scale bars: 20μm). Center panel: Endothelial cell marker CD31 was stained (green) in a radial CNV (white arrowheads; scale bars: 20μm). Nuclei were stained with DAPI (blue). Right panel: Elastic-van Gieson (EVG) staining of Bruch's membrane (BM: red dotted frame). *HtrA1* Tg mice displayed ruptures and deficiencies in the BM (scale bars: 10μm).

The promoter region of *HTRA1* is highly associated with CNV in AMD patients. Risk variant leads to a 2.7-fold increase in *HTRA1* mRNA in the RPE and confers an estimated population attributable risk of 49.3%. In our mouse model, a 98.2-fold increase in *HtrA1* mRNA in the mouse eye led to an attributed CNV risk of 18.2%. Our results suggest that the *HTRA1* expression level may influence CNV progression, and that other unknown environmental risk factors may influence the risk variant. As none of the WT mice displayed CNV, the *HTRA1* expression level seems to be more influential than aging on CNV development. Previous studies have suggested that overexpression of *HTRA1* in the RPE can lead to alterations of the Bruch's membranes, with fragmentation of the elastic lamina [26, 27]. Similarly, we found that ubiquitous overexpression of *HTRA1* resulted in ruptured or deficient Bruch's membranes.

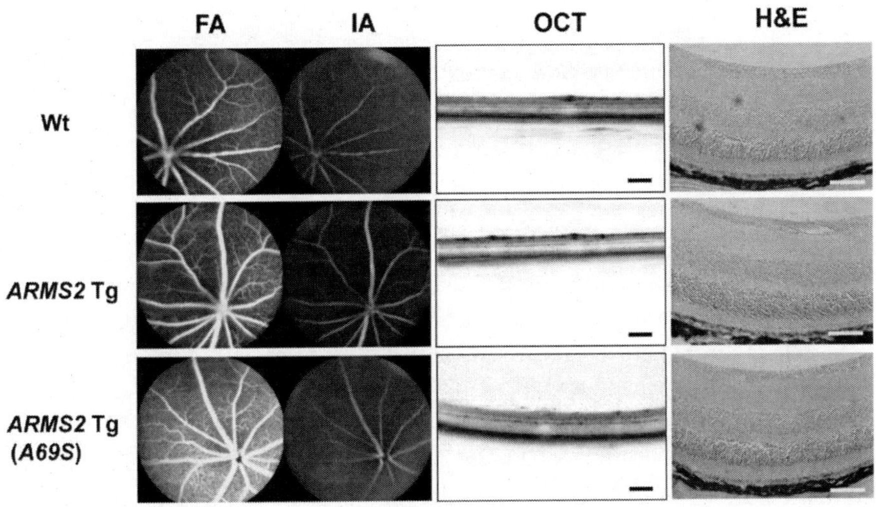

Figure 5. No retinal abnormalities in 12-month-old *ARMS2* Tg and *ARMS2* (A69S) Tg mice. *ARMS2* Tg and *ARMS2* (A69S) Tg mice showed no retinal abnormalities at 1 year by fundus observation (FA: fluorescein angiography, IA: indocyanine green angiography, OCT: optical coherence tomography, scale bars: 200μm) or hematoxylin and eosin (H & E) staining (40× magnification; scale bars: 20μm).

On the other hand, previous reports have suggested that *ARMS2* is a constituent of the ECM. This localization suggests that *ARMS2* may be necessary for proper matrix function and may protect against drusen formation [17]. We examined 20 aged *ARMS2* Tg mice and 20 aged *ARMS2* (A69S) Tg mice, none of which displayed any retinal changes. These results demonstrate that ubiquitous overexpression of *ARMS2* or *ARMS2* (A69S) does not lead to typical AMD phenotypes, such as drusen or CNV formation (Figure 5) [30]. We were unable to confirm whether *ARMS2* protects against CNV formation.

4. Clinical Importance of *HTRA1*

The formation of new vessels from the choroid into the subretinal space is the most important change that predisposes patients with druse to macular detachment and irreversible loss of central vision. Examinations to identify blood vessel abnormalities include a visual test and a fundus angiogram using fluorescein or indocyanine green. However, the clinical changes of early subretinal neovascularization are subtle and may be easily overlooked. Therefore, when a patient with evidence of AMD has sudden or recent central vision loss, ophthalmologists must suspect the presence of subretinal neovascularization. Unfortunately, disease progression frequently occurs very quickly, such that it is too late to intervene when the disease is discovered. In these cases, the prevention of AMD becomes very important.

Our results demonstrated that ubiquitous expression of *HtrA1* in aged mice could induce the development of subretinal neovascularization similar to human AMD patients. This possibility suggests that *HtrA1* overexpression alone is a strong risk factor for wet AMD. As a first step to prevent wet AMD, our findings may be useful for developing a genetic test to identify, at an early stage, whether a patient will suffer from this disease in the future.

Upregulation of vascular endothelial growth factor (VEGF) and degradation of Bruch's membrane are risk factors for CNV [31, 32), both of which may be affected by *HTRA1* overexpression. VEGF plays a key role in the pathogeneses of exudative AMD and ocular neovascularization in general [33]. Bruch's membrane, which relies on the RPE and choroid, is an important factor in AMD development, particularly CNV formation in wet AMD. As a serine protease involved in the degradation of ECM proteins, such as fibronectin and aggrecan [34, 35], *HTRA1* overexpression in the RPE increases fibronectin degradation [26]. The resulting fibronectin fragments have proteolytic activity. Furthermore, these fragments stimulate the release of matrix metalloproteinases, which might further contribute to ECM destabilization [36], and cytokines (e.g., tumor necrosis factor [TNF]-α), which upregulate VEGF secretion in RPE cells [37]. We found that *HtrA1* overexpression resulted in ruptures and deficiencies in Bruch's membrane.

In conclusion, our *HtrA1* Tg mouse model of wet AMD may be used to understand the pathogenic mechanism of wet AMD and may contribute to the development of new drug treatments. Treating wet AMD may involve anti-VEGF (e.g., Bevacizumab, Ranibizumab, or Aflibercept), thermal laser, and photodynamic therapies. Treating wet AMD can generally reduce, but not eliminate, the risk of severe vision loss; the longer period of time that abnormal vessels leak or grow, the greater the risk of vision loss becomes. Additionally, if abnormal blood vessel growth happens in one eye, there is an increased risk that it will also occur in the other. Patients have best chance of preserving central vision when wet AMD is diagnosed and treated at an early stage. Thus, early detection by genetic testing will be critical in advancing treatment for AMD.

References

[1] Ding, X., Patel, M. and Chan, C. C. (2009) Molecular pathology of age-related macular degeneration. *Progress in retinal and eye research*, 28, 1-18.

[2] Friedman, D. S., O'Colmain, B. J., Munoz, B., Tomany, S. C., McCarty, C., de Jong, P. T., Nemesure, B., Mitchell, P. and Kempen, J. (2004) Prevalence of age-related macular degeneration in the United States. *Archives of ophthalmology*, 122, 564-572.

[3] Chen, Y., Bedell, M. and Zhang, K. (2010) Age-related macular degeneration: genetic and environmental factors of disease. *Mol Interv*, 10, 271-281.

[4] Mitchell, P., Smith, W. and Wang, J. J. (1998) Iris color, skin sun sensitivity, and age-related maculopathy. The Blue Mountains Eye Study. *Ophthalmology*, 105, 1359-1363.

[5] Sperduto, R. D. and Hiller, R. (1986) Systemic hypertension and age-related maculopathy in the Framingham Study. *Archives of ophthalmology*, 104, 216-219.

[6] Klein, R., Klein, B. E., Linton, K. L. and DeMets, D. L. (1993) The Beaver Dam Eye Study: the relation of age-related maculopathy to smoking. *Am J Epidemiol*, 137, 190-200.

[7] Ritter, L. L., Klein, R., Klein, B. E., Mares-Perlman, J. A. and Jensen, S.C. (1995) Alcohol use and age-related maculopathy in the Beaver Dam Eye Study. *American journal of ophthalmology*, 120, 190-196.

[8] Kawasaki, R., Wang, J. J., Ji, G. J., Taylor, B., Oizumi, T., Daimon, M., Kato, T., Kawata, S., Kayama, T., Tano, Y. et al. (2008) Prevalence and risk factors for age-related macular degeneration in an adult Japanese population: the Funagata study. *Ophthalmology*, 115, 1376-1381, 1381 e1371-1372.

[9] Goto, A., Akahori, M., Okamoto, H., Minami, M., Terauchi, N., Haruhata, Y., Obazawa, M., Noda, T., Honda, M., Mizota, A. et al. (2009) Genetic analysis of typical wet-type age-related macular degeneration and polypoidal choroidal vasculopathy in Japanese population. *J Ocul Biol Dis Infor*, 2, 164-175.

[10] Fritsche, L. G., Chen, W., Schu, M., Yaspan, B. L., Yu, Y., Thorleifsson, G., Zack, D. J., Arakawa, S., Cipriani, V., Ripke, S. et al. (2013) Seven new loci associated with age-related macular degeneration. *Nat Genet*, 45, 433-439, 439e431-432.

[11] Klein, R. J., Zeiss, C., Chew, E. Y., Tsai, J. Y., Sackler, R. S., Haynes, C., Henning, A. K., SanGiovanni, J. P., Mane, S.M., Mayne, S. T. et al. (2005) Complement factor H polymorphism in age-related macular degeneration. *Science*, 308, 385-389.

[12] Hageman, G. S., Anderson, D. H., Johnson, L. V., Hancox, L. S., Taiber, A. J., Hardisty, L. I., Hageman, J. L., Stockman, H. A., Borchardt, J. D., Gehrs, K. M. et al. (2005) A common haplotype in the complement regulatory gene factor H (HF1/CFH) predisposes individuals to age-related macular degeneration. *Proceedings of the National Academy of Sciences of the United States of America*, 102, 7227-7232.

[13] Okamoto, H., Umeda, S., Obazawa, M., Minami, M., Noda, T., Mizota, A., Honda, M., Tanaka, M., Koyama, R., Takagi, I. et al. (2006) Complement factor H polymorphisms in Japanese population with age-related macular degeneration. *Mol Vis*, 12, 156-158.

[14] Kondo, N., Honda, S., Ishibashi, K., Tsukahara, Y. and Negi, A. (2007) LOC387715/*HTRA1* variants in polypoidal choroidal vasculopathy and age-related macular degeneration in a Japanese population. *American journal of ophthalmology*, 144, 608-612.

[15] Fritsche, L. G., Loenhardt, T., Janssen, A., Fisher, S. A., Rivera, A., Keilhauer, C. N. and Weber, B. H. (2008) Age-related macular degeneration is associated with an unstable *ARMS2* (LOC387715) mRNA. *Nat Genet*, 40, 892-896.

[16] Kanda, A., Chen, W., Othman, M., Branham, K. E., Brooks, M., Khanna, R., He, S., Lyons, R., Abecasis, G. R. and Swaroop, A. (2007) A variant of mitochondrial protein LOC387715/*ARMS2*, not *HTRA1*, is

strongly associated with age-related macular degeneration. *Proceedings of the National Academy of Sciences of the United States of America*, 104, 16227-16232.

[17] Kortvely, E., Hauck, S. M., Duetsch, G., Gloeckner, C. J., Kremmer, E., Alge-Priglinger, C. S., Deeg, C. A. and Ueffing, M. (2010) *ARMS2* is a constituent of the extracellular matrix providing a link between familial and sporadic age-related macular degenerations. *Investigative ophthalmology & visual science*, 51, 79-88.

[18] Shiga, A., Nozaki, H., Yokoseki, A., Nihonmatsu, M., Kawata, H., Kato, T., Koyama, A., Arima, K., Ikeda, M., Katada, S. et al. (2011) Cerebral small-vessel disease protein *HTRA1* controls the amount of TGF-beta1 via cleavage of proTGF-beta1. *Hum Mol Genet*, 20, 1800-1810.

[19] Zhang, L., Lim, S. L., Du, H., Zhang, M., Kozak, I., Hannum, G., Wang, X., Ouyang, H., Hughes, G., Zhao, L. et al. (2012) High temperature requirement factor A1 (*HTRA1*) gene regulates angiogenesis through transforming growth factor-beta family member growth differentiation factor 6. *The Journal of biological chemistry*, 287, 1520-1526.

[20] Jiang, J., Huang, L., Yu, W., Wu, X., Zhou, P. and Li, X. (2012) Overexpression of *HTRA1* leads to down-regulation of fibronectin and functional changes in RF/6A cells and HUVECs. *PloS one*, 7, e46115.

[21] Ferrer-Vaquer, A., Maurey, P., Werzowa, J., Firnberg, N., Leibbrandt, A. and Neubuser, A. (2008) Expression and regulation of *HTRA1* during chick and early mouse development. *Dev Dyn*, 237, 1893-1900.

[22] Chien, J., Ota, T., Aletti, G., Shridhar, R., Boccellino, M., Quagliuolo, L., Baldi, A. and Shridhar, V. (2009) Serine protease *HTRA1* associates with microtubules and inhibits cell migration. *Mol Cell Biol*, 29, 4177-4187.

[23] Hara, K., Shiga, A., Fukutake, T., Nozaki, H., Miyashita, A., Yokoseki, A., Kawata, H., Koyama, A., Arima, K., Takahashi, T. et al. (2009) Association of *HTRA1* mutations and familial ischemic cerebral small-vessel disease. *The New England journal of medicine*, 360, 1729-1739.

[24] Dewan, A., Liu, M., Hartman, S., Zhang, S. S., Liu, D. T., Zhao, C., Tam, P. O., Chan, W. M., Lam, D. S., Snyder, M. et al. (2006) *HTRA1* promoter polymorphism in wet age-related macular degeneration. *Science*, 314, 989-992.

[25] Yang, Z., Tong, Z., Chen, Y., Zeng, J., Lu, F., Sun, X., Zhao, C., Wang, K., Davey, L., Chen, H. et al. (2010) Genetic and functional dissection of *HTRA1* and LOC387715 in age-related macular degeneration. *PLoS Genet*, 6, e1000836.

[26] Vierkotten, S., Muether, P. S. and Fauser, S. (2011) Overexpression of *HTRA1* leads to ultrastructural changes in the elastic layer of Bruch's membrane via cleavage of extracellular matrix components. *PloS one*, 6, e22959.

[27] Jones, A., Kumar, S., Zhang, N., Tong, Z., Yang, J.H., Watt, C., Anderson, J., Amrita, Fillerup, H., McCloskey, M. et al. (2011) Increased expression of multifunctional serine protease, *HTRA1*, in retinal pigment epithelium induces polypoidal choroidal vasculopathy in mice. *Proceedings of the National Academy of Sciences of the United States of America*, 108, 14578-14583.

[28] Yang, Z., Camp, N. J., Sun, H., Tong, Z., Gibbs, D., Cameron, D. J., Chen, H., Zhao, Y., Pearson, E., Li, X. et al. (2006) A variant of the *HTRA1* gene increases susceptibility to age-related macular degeneration. *Science*, 314, 992-993.

[29] Iejima, D., Itabashi, T., Kawamura, Y., Noda, T., Yuasa, S., Fukuda, K., Oka, C. and Iwata, T. (2015) *HTRA1* (High Temperature Requirement A Serine Peptidase 1) Gene Is Transcriptionally Regulated by Insertion/Deletion Nucleotides Located at the 3' End of the *ARMS2* (Age-related Maculopathy Susceptibility 2) Gene in Patients with Age-related Macular Degeneration. *The Journal of biological chemistry*, 290, 2784-2797.

[30] Nakayama, M., Iejima, D., Akahori, M., Kamei, J., Goto, A. and Iwata, T. (2014) Overexpression of *HTRA1* and exposure to mainstream cigarette smoke leads to choroidal neovascularization and subretinal deposits in aged mice. *Investigative ophthalmology & visual science*, 55, 6514-6523.

[31] Nakashizuka, H., Mitsumata, M., Okisaka, S., Shimada, H., Kawamura, A., Mori, R. and Yuzawa, M. (2008) Clinicopathologic findings in polypoidal choroidal vasculopathy. *Investigative ophthalmology & visual science*, 49, 4729-4737.

[32] Witmer, A. N., Vrensen, G. F., Van Noorden, C. J. and Schlingemann, R. O. (2003) Vascular endothelial growth factors and angiogenesis in eye disease. *Progress in retinal and eye research*, 22, 1-29.

[33] Ng, E. W. and Adamis, A. P. (2005) Targeting angiogenesis, the underlying disorder in neovascular age-related macular degeneration. *Canadian journal of ophthalmology. Journal canadien d'ophtalmologie*, 40, 352-368.

[34] Grau, S., Richards, P. J., Kerr, B., Hughes, C., Caterson, B., Williams, A. S., Junker, U., Jones, S. A., Clausen, T. and Ehrmann, M. (2006) The

role of human *HTRA1* in arthritic disease. *The Journal of biological chemistry*, 281, 6124-6129.

[35] Chamberland, A., Wang, E., Jones, A. R., Collins-Racie, L. A., La Vallie, E. R., Huang, Y., Liu, L., Morris, E. A., Flannery, C. R. and Yang, Z. (2009) Identification of a novel *HTRA1*-susceptible cleavage site in human aggrecan: evidence for the involvement of *HTRA1* in aggrecan proteolysis *in vivo*. *The Journal of biological chemistry*, 284, 27352-27359.

[36] Lambert Vidmar, S., Lottspeich, F., Emod, I., Imhoff, J. M. and Keil-Dlouha, V. (1991) Collagen-binding domain of human plasma fibronectin contains a latent type-IV collagenase. *European journal of biochemistry/FEBS*, 201, 79-84.

[37] Terasaki, H., Kase, S., Shirasawa, M., Otsuka, H., Hisatomi, T., Sonoda, S., Ishida, S., Ishibashi, T. and Sakamoto, T. (2013) TNF-alpha decreases VEGF secretion in highly polarized RPE cells but increases it in non-polarized RPE cells related to crosstalk between JNK and NF-kappaB pathways. *PloS one*, 8, e69994.

In: Age-Related Macular Degeneration
Editor: Melanie Carter
ISBN: 978-1-63483-329-5

Chapter III

The New Era of Omega-3 Fatty Acids Supplementation: Therapeutic Effects on Dry Age-Related Macular Degeneration

Tassos Georgiou[1,*] and Ekatherine Prokopiou[2]
[1]Ophthalmologist and Head of Ophthalmos Research and Educational Institute, Nicosia, Cyprus
[2]Post-doctoral Research Scientist at Ophthalmos Research and Educational Institute, Nicosia, Cyprus

Abstract

One of the major causes of reduced vision over the age of 50 is age-related macular degeneration (AMD). Although such a common pathology, there are no current guidelines for the first-line treatment of dry AMD. The aim of this study is to evaluate the therapeutic effects of high omega-3 fatty acids as anti-inflammatory agents in two sub-groups of dry AMD patients with 1) mild to moderate visual impairment and 2)

[*] Authors' contact information: Ophthalmos Research and Educational Institute, Morfou 48, Engomi, Nicosia, 2417, Cyprus, Tel: + 357 22 464344, Fax + 357 22 464345, Email: tassosgeorgiou@hotmail.com, prokopiou.katerina@ucy.ac.cy.

with severe visual impairment (blindness). The key feature of this investigation is the frequent monitoring of the levels of specific fatty acids in patient's blood in order to adjust the treatment dose within the ideal therapeutic window.

Following the positive outcome from our initial observational studies in patients with dry AMD, who demonstrated significant improvement in visual acuity (gain of ≥ 1 line of vision in 4.5 months) when taking a total of 5 g/day eicosapentaenoic acid (EPA) and docosahexaenoic acid (DHA), additional studies were encouraged. The latest data which is presented in this chapter suggests that the eyes which had the greatest gain in vision (≥ 15 letters gain at 6 months) were from patients with mild to moderate visual impairment, who were taking between 5-7.5 g/day EPA and DHA and had a ratio of arachidonic acid (AA)/EPA < 2. In addition, a sub-group of dry AMD patients with severe visual impairment (< 6/60), showed significant increase in their visual acuity only 3 months following treatment with omega-3 fatty acids. The preliminary results indicate a promising therapeutic regime for dry AMD and perhaps for other types of retinopathies as well.

Although initial results are encouraging, further investigations are necessary to establish a better understanding of the mode of action of these supplements and to observe their long-term effects.

Keywords: age-related macular degeneration, inflammation, omega-3 fatty acids, arachidonic acid, eicosapentaenoic acid, visual acuity, blindness

Introduction

Age-related macular degeneration (AMD) is the leading cause of visual impairment (VI) and blindness in the elderly and it is estimated that 200 million people suffer worldwide. By the year 2040, the number of these individuals is estimated to be increased by 50% [1, 2]. VI is defined as a limitation in one or more functions of the eye or visual system, most commonly impairment of visual acuity (VA), visual fields, contrast sensitivity and colour vision. According to the Access Economics for AMD Alliance International, the definitions of visual acuity for mild VI, moderate VI and blindness are as following:

- Mild VI is defined as best-corrected VA less than 6/12 but ≥ 6/18 in the better-seeing eye.

- Moderate VI is defined as best-corrected VA less than 6/18 but ≥ 6/60 in the better-seeing eye.
- Blindness is defined as best-corrected VA less than 6/60 in the better-seeing eye.

AMD was identified as causing 9% of all global VI in the year 2002 [3], although the proportion is substantially higher (up to 50%) in developed WHO sub-regions. There is an estimated increase of 13 million AMD global cases by the year 2020 and is associated with direct rise in health care system expenditure, deadweight welfare loss and informal care costs. In 2010 there was a direct health care system expenditure of $255 billion associated with AMD and this number is estimated to increase up to $294 billion by the year 2020. Taking into consideration the significant impact of AMD on the quality of life of patients and their carers, in addition to the global economic and health burden of this pathology, our research team has suggested a possible therapeutic regime using supplementation with omega-3 polyunsaturated fatty acids (PUFAs). Therefore, this chapter familiarises with the pathogenesis of AMD before presenting some of the most important pre-clinical and clinical studies conducted with omega-3 PUFAs, with particular emphasis given on their anti-inflammatory effects. Preliminary results from our observational studies are presented and directions regarding how to benefit from the omega-3 PUFAs are discussed.

Age-Related Macular Degeneration

The main symptoms of early AMD is blurring of central vision, the need of brighter light to read by, colour appearance is less bright and difficulty in recognising faces. There are two different types of AMD, the dry form which is the most common one and occurs in 9 out of 10 cases, and the wet form (choroid neovascularisation, CNV) which occurs in about 1 in 10 cases. In dry AMD, the retinal pigment epithelial (RPE) cells of the macula, which are crucial for the function of the rods and cones, will gradually degenerate. In wet AMD, in addition to the RPE cells' degeneration, newly formed blood vessels grow from the choroid, break through Bruch's membrane and pass into the macula part of the retina. These vessels are immature and leak fluid within the retina resulting in scarring of the macula and damage of the rods and cones [4, 5].

Pathogenesis of Age-Related Macular Degeneration

Currently, there is no definite cause for the pathogenesis of AMD, but several different aetiologies. Aging is one of the most common contributing factors of AMD, due to the accumulation of oxidised lipoproteins and free radicals in the retina and choroid. This in turn, results in oxidative stress and a decrease in the number of RPE cells and photoreceptors [6, 7]. In addition to oxidative stress, natural age advancement also leads to immunosenescence which is the gradual deterioration of the immune system, specifically the T-cell population [8, 9].

Genetic predisposition, as with multiple pathologies, plays a role in the development of AMD; although there are several genetic associations, the most studied ones are some polymorphic links, in particular related to inflammatory genes, such as the *complement factor H* (*CFH*) and some complement components (e.g., *C3* and *C2*) [10, 11]. The *CFH* gene which controls the activation of the complement system through the alternative pathway was found to be associated with an increased risk of developing AMD [12, 13]. Environmental factors, including smoking, sunlight exposure, high-fat diet, obesity and diabetes, are all associated with the development and progression of AMD [14].

Para-Inflammation in Age-Related Macular Degeneration

Furthermore, a tissue adaptive response, recently described as para-inflammation, where the innate immune system mount a low-grade inflammatory response in order to restore tissue homeostasis, has been implicated in the pathogenesis of AMD [15, 16]. In particular, chronic inflammatory infiltrates, such as macrophages, lymphocytes and mast cells, have been found in the choroid of AMD eye donors [17, 18]. Inflammatory-related proteins, including C-reactive protein (CRP) (19-22), interleukin-6 (IL-6) [22, 23] and tumour necrosis factor-α (TNF-α) [22, 24], have been shown to be associated with AMD; however, the results from different groups are inconsistent.

The fact that systemic inflammatory markers are not strongly related with AMD might suggest that local low-grade inflammation is more likely to be involved in its pathogenesis. In addition, resident retinal microglia were found to be activated in the outer nuclear layer, in regions of ongoing rod cell death, in patients with AMD, retinitis pigmentosa (RP) and late-onset retinal degeneration [25]. The microglia cell activation is an indication of an immune response to ocular injury or inflammation, as well as retinal degeneration.

During normal aging as well as in pathological conditions, such as AMD, there is an observed accumulation of microglia in the subretinal space, localised in the areas of RPE cell death. Apart from microglia activation, complement activation is also involved in aging and in both forms of AMD. It is suggested that the damage of RPE cells and photoreceptors in AMD may, at least in part, be caused directly by complement activation at the retinal/choroidal interface [26, 27]. Evidence demonstrated that a local complement regulatory system exists in the eye, by the detection of complement components, such as the C3, membrane cofactor protein (MCP), decay-acceleration factor (DAF), membrane inhibitor of reactive lysis (CD59) and cell surface regulator of complement [28, 29].

Omega-3 Polyunsaturated Fatty Acids

Over the last few decades there is an increasing interest in the role of omega-3 PUFAs and inflammation in general. Numerous evidence exist from pre-clinical and clinical studies which prove the effectiveness of omega-3 PUFAs against heart disease, cancer, diabetes, neurological and autoimmune diseases [30].

Currently, several studies have been focusing on the therapeutic role of omega-3 PUFAs, which are considered anti-inflammatory molecules. The resolution of inflammation is an active process primarily driven by a new family of mediators, termed resolvins, derived from the omega-3 PUFAs, eicosapentaenoic acid (EPA, C20:5ω-3) and docosahexaenoic acid (DHA, C22:6ω-3) [31]. These PUFAs are highly concentrated in the brain and retina and have an important role in the neuronal development and damage repair [32]. DHA is abundantly expressed in the photoreceptors and vital retinal functions depend on its existence [33].

Among the major mediators of the inflammatory response is the generation of pro-inflammatory eicosanoids generated from the omega-6 PUFA, arachidonic acid (AA, C20:4ω-6). These include pro-inflammatory prostaglandins (e.g., PGE_2) and leukotrienes (e.g., LTB_4), which can act as mediators for leukocyte chemotaxis and inflammatory cytokine production. The balance between the pro- and anti-inflammatory molecules plays a key role in the disease progression and the resolution of an inflammatory response.

Treatment Options for Age-Related Macular Degeneration

With regards to the wet AMD, treatment options are based on anti-VEGF (vascular endothelial growth factor) therapies which aim to attenuate angiogenesis and vascular permeability. However, this type of targeted therapy does not lead to complete vascular or disease regression [34]. To assess the long-term outcomes, Rofagha et al., investigated the effect of intensive ranibizumab therapy in exudative AMD patients 7 to 8 years after initiation of the treatment. One fifth of the patients demonstrated good visual outcomes 7 years post-treatment, whereas another third had poor outcomes. Almost half of the eyes were stable, whereas one third declined by 15 letters or more [35]. These results may indicate that even a long-term therapeutic regime does not reduce the risk for substantial visual decline.

In contrast, for the dry form of AMD there are no current guidelines for the first-line treatment, although several anti-oxidants, vitamins and zinc may reduce its progression according to the Age-related Eye Disease Study (AREDS). This study was a major clinical trial sponsored by the National Eye Institute which was designed to evaluate the effect of high doses of vitamin C (500 mg), vitamin E (400 international units), β-carotene (15 mg), zinc (80 mg) and copper (2 mg) on the progression of AMD and cataract [36]. Following the AREDS, an additional study was performed, the AREDS2, which was a multi-centre five-year randomised trial, designed to examine the effects of oral supplementation of macular xanthophylls (10 mg lutein and 2 mg zeaxanthin) and/or omega-3 PUFAs (DHA and EPA, 1 g) on the progression to advanced AMD. Overall, there was no additional benefit from adding the omega-3 PUFAs or a mixture of lutein and zeaxanthin to the formulation. Although the addition of omega-3 to the AREDS formulation was not proven beneficial, it is believed that higher doses of EPA and DHA may have a desirable effect. Therefore, further studies were conducted or are still ongoing in order to investigate the possible mechanisms of action of the omega-3 PUFAs or to examine any positive outcomes on the disease progression.

In addition, a current review by Buschini et al., summarised the most recent developments for the management of dry AMD, including ongoing trials with anti-β-amyloid, anti-C5 and anti-factor D antibodies and other anti-inflammatory agents such as fluocinolone acetonide and rapamycin. New strategies aim to reduce drusen formation, lower the accumulation of toxic by-products from the visual cycle, improve the choroidal perfusion and develop targeted gene therapy [37].

Current Research in Age-Related Macular Degeneration

Numerous *in vitro* studies demonstrated that treatment of endothelial cells with omega-3 PUFAs effectively inhibited pro-inflammatory responses through modulation of nuclear factor-κB (NF-κB), TNF-α and interleukin-1β (IL-1β)-induced cell adhesion molecule (CAM) expression [38-41].

Dry Age-Related Macular Degeneration

Furthermore, several animal studies focused on the dietary supplementation of omega-3 PUFAs in murine models for macular degeneration. In particular, Tuo et al., reported the therapeutic effect of a high omega-3 diet in the $Ccl2^{-/-}/Cx3cr1^{-/-}$ double knock out model, which develops focal retinal lesions with certain features of AMD [42]. The high omega-3 diet included 1.9 wt % of each EPA and DHA, 0.66 wt % α-linolenic acid (α-LNA) and 0.4 wt % docosapentaenoic acid (DPA). The omega-6/omega-3 ratio was 2.9, where the omega-6 source was from the linoleic acid (LA) only. Specifically, animals that ingested a high omega-3 diet for up to 8 months of age, showed progression of retinal lesions compared with the low omega-3 diet group. This effect was suggested to be through a reduction in the AA metabolism, as demonstrated by the decreased pro-inflammatory derivatives (PGE_2 and LTB_4). High levels of dietary omega-3 PUFAs may result in the incorporation of EPA into cell membrane phospholipids at the expense of AA, leading to less substrate available for eicosanoid synthesis [43]. In contrast with PGE_2, higher serum levels of PGD_2 (an anti-inflammatory mediator [44]), was observed in the high omega-3 PUFAs group, indicating a protective effect against inflammation. In addition, there was lower ocular *TNF-α* and *IL-6* transcript levels in the high omega-3 group, suggesting that reactive mediators of omega-3 PUFAs may also regulate differential gene expression [42].

Following treatment with AREDS2 formulation, Ramkumar et al., reported its effect on the $Ccl2^{-/-}/Cx3cr1^{-/-}$ model of AMD [45]. This formulation included high doses of omega-3 PUFAs (54.9 mmol EPA/kg diet and 25.2 mmol DHA/kg diet), 17.6 mmol lutein/kg diet and 1.76 mmol zeaxanthin/kg diet. After 3 months of treatment, the animals fed with the high omega-3 PUFAs diet showed significant lesion regression, following fundoscopic examination and a great reduction in the ocular A2E concentration (a fluorophore found in lipofuscin and RPE phagolysosomes). Morphological changes were noted, where the outer nuclear layer thickness was greater in the high omega-3 treated group than that in the control one. The retinal expression of pro-inflammatory mediators, including *TNF-α*, *Cyclooxygenase-2* (*Cox-2*), *IL-1β*, *VEGF* and *inducible nitric oxide synthase* (*iNos*) was much lower in the high omega-3 treated group compared with the control. The AA concentration in the retina was found to be lower and the EPA higher in the high omega-3 group compared to the control, where the retina AA/EPA ratio was estimated to be 2.26. However, the serum concentration of PGE_2 did not significantly differ between the two groups [45]. This fact may indicate that the concentration of fatty acids (and their metabolites) in the serum and retina is not directly correlated.

Wet Age-Related Macular Degeneration

Additional animal studies have been performed using models related to wet AMD or to other ocular pathologies, including retinal angiogenesis or diabetic retinopathy. Yanai et al., demonstrated that metabolites derived from the omega-3 PUFAs, which are generated through the Cytochrome P450 enzymes (CYP450), are potent inhibitors of intraocular neovascular disease, such as wet AMD [46]. It is known that the omega-6 PUFAs, including AA, generate CYP metabolites, the epoxyeicosatrienoic acids (EETs), which are associated with the VEGF-activated signalling cascade, leading to angiogenesis [47]. On the other hand, the CYP-generated metabolites of EPA and DHA, namely 17,18-epoxyeicosatetraenoic acid (17,18-EEQ) and 19,20-epoxydocosapentaenoic acid (19,20-EDP), respectively, have shown anti-angiogenic properties [48]. Therefore, Yanai et al., investigated the dietary enrichment with omega-3 PUFAs in a mouse model of laser-induced CNV and demonstrated suppression of CNV (possibly through increased expression of peroxisome proliferator-activated receptor-γ, PPAR-γ), vascular leakage, immune cell recruitment and adhesion molecules (E-selectin and intracellular adhesion molecule-1, Icam-1) to the lesion site. In addition, *VEGF* expression was suppressed in the retina and choroid of the mice fed with the high omega-3 diet. A significant increase in the serum levels of anti-inflammatory eicosanoids was observed in the high omega-3 group, which was mediated through the CYP-metabolites, 17,18-EEQ and 19,20-EDP [46].

Similar studies by Connor et al., were previously performed, evaluating the therapeutic effect of omega-3 PUFAs on hypoxia-induced pathological neovascularisation in a mouse model of oxygen-induced retinopathy [49]. The results suggested that by increasing the level of omega-3 PUFAs either by dietary or genetic means (using *Fat-1* transgenic model which converts omega-6 to omega-3 PUFAs) there was a reduced hypoxic stimulus for neovascularisation. This effect was mediated through the bioactive metabolites neuroprotectinD1, resolvinD1 and resolvinE1, through reduction in the TNF-α expression, which was found to be present in a subset of microglia within the retinal vessels [49].

Furthermore, in a clinical setting Renzende et al., examined the effect of omega-3 PUFAs supplementation (1052 mg fish oil, 600 mg EPA/DHA) on the levels of vitreous VEGFA, in patients with wet AMD who were receiving intravitreal anti-VEGF therapy. Interestingly, the group of patients that was supplemented with omega-3 showed lower levels of VEGFA in the vitreous but similar levels in the plasma compared to the other groups [50].

Observational Studies in Dry AMD Patients

In addition to the preclinical studies, an open-label pilot study performed by Georgiou et al., investigated the therapeutic effect of controlled doses of high omega-3 PUFAs in patients with dry AMD. The supplement formulation included 3.4 g EPA and 1.6 g DHA, where patients followed this treatment on a daily basis for 6 months. Significant improvement in visual acuity was observed in all patients with dry AMD, and by 4.5 months of treatment all patients had gained a minimum 1 line of vision consisting of 5 letters [51].

Further observational studies took place with 61 eyes from 49 patients with dry AMD, from both sexes and mean age of 72.5 ± 0.98. The VA of these patients was rated from mild to moderate (6/9 to 6/60). Patients were supplemented with 5-7.5 g of an omega-3 formulation divided in two daily doses. The omega-3 concentrates consisted of purified ethyl esters rich in EPA (400 mg) and DHA (200 mg) per gram of the liquid formulation. The dosage used in this pilot study provided approximately 5 g of EPA and 2.5 g of DHA per day. Patients were asked to continue with their current diet as normal, and not to increase their consumption of food containing essential fatty acids. In case where patients were on anti-coagulation drugs or suffered from other conditions that would interfere with this therapeutic regime, they were not included in the study.

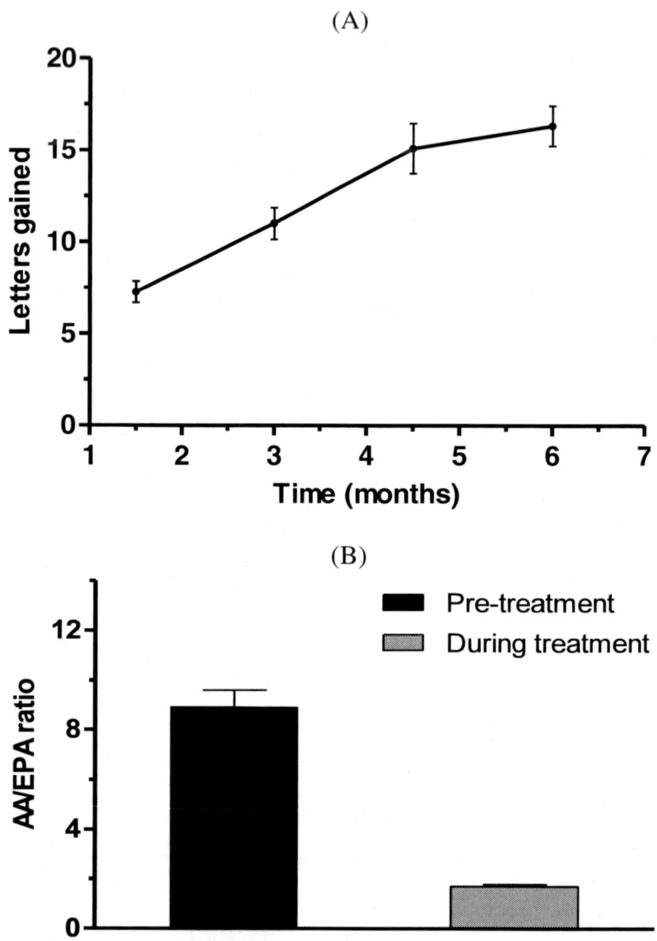

Figure 1. The effect of omega-3 supplementation on visual acuity in patients with dry AMD according to the letters gained (A) and the change in AA/EPA ratio (B), mean ± SEM.

Follow up was performed at 1.5, 3, 4.5 and 6 months. VA was examined at each follow up using the EDTRS chart and the blood AA/EPA ratio was measured prior and during treatment using a gas chromatography technique. The optimum therapeutic AA/EPA ratio is considered to be < 2. The mean gain of letters ± SEM at 1.5, 3, 4.5 and 6 months was 7.3 ± 0.6, 11.2 ± 0.86, 15.1 ± 0.4 and 16.3 ± 1.1, respectively (Figure 1A).

The mean AA/EPA blood ratio prior to treatment was 8.9 ± 0.7, whereas during the period of treatment was reduced to 1.7 ± 0.08 (Figure 1B). No side effects were reported by any of the patients involved in this study. The closer the AA/EPA to 1 the more pronounced the effect was, with regards to the number of letters gained.

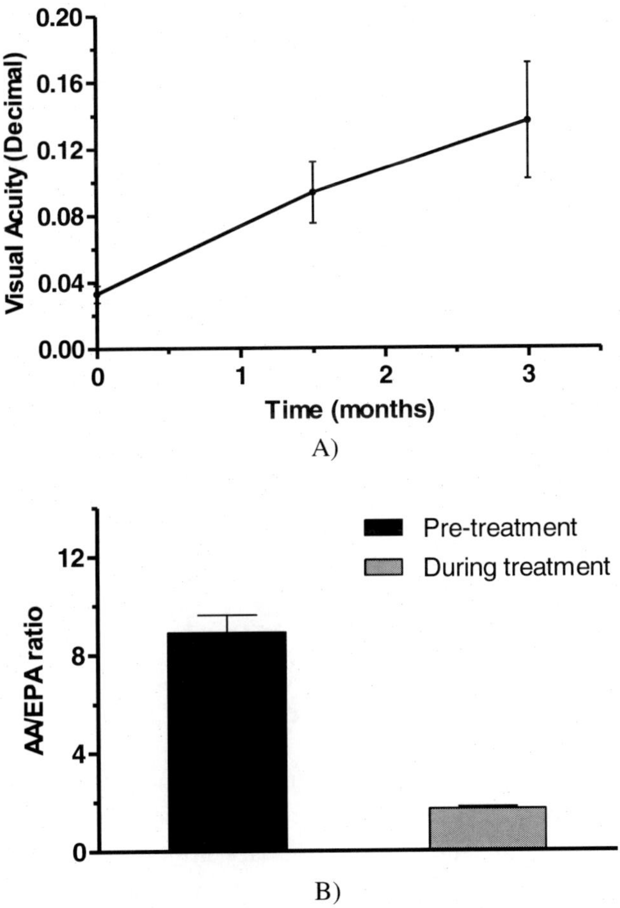

Figure 2. The effect of omega-3 supplementation on visual acuity in Decimals in patients with dry AMD with initial VA < 6/60 (A) and the change in AA/EPA ratio (B), mean ± SEM.

Furthermore, a different sub-group of patients with VA < 6/60, which according to the Access Economics for AMD Alliance International fall into the blindness category, participated in a similar study. Twelve eyes from nine patients with dry AMD (VA < 6/60), from both sexes and mean age of 78.3 ± 1.7 were supplemented with the omega-3 formulation (5-7.5 g daily). Follow up on VA was performed at 1.5 and 3 months, along with the AA/EPA blood monitoring (Figure 2A and 2B).

The mean AA/EPA blood ratio prior to treatment was 7.9 ± 0.9, whereas during the period of treatment was reduced to 1.9 ± 0.5. The mean initial vision in Decimals ± SEM was 0.033 ± 0.004, whereas following 1.5 months of treatment with omega-3 there was a significant improvement in vision at 0.09 ± 0.01. At 3 months of treatment, there was an even greater increase (> 4 times) in VA at 0.14 ± 0.03 which in theory, but also in practise, removes those patients from the blindness category. No side effects were reported by any of the patients involved in this study. The study is still ongoing in order to investigate the long-term effect of omega-3 supplementation.

Although these observations are only preliminary results with small number of participants, are very promising and encourage further studies to take place in order to clarify the beneficial effect of the omega-3 fatty acids in certain situations.

In summary, evidence from the literature and from our observational studies imply that supplementation of high doses of omega-3 PUFAs has a positive effect on the disease progression, including both dry and wet AMD, by reducing the inflammatory response through active mediators and metabolites.

Conclusion

Chronic low grade inflammation attacks the body's organs for years until enough damage occurs to produce a chronic disease. It is inflammation below the threshold of pain which is not initially noticeable until chronic organ damage develops. As for the eyes, vision impairment is the first obvious symptom. Anti-inflammatory drugs, such as steroids and non-steroidal anti-inflammatory drugs could reduce chronic inflammation; although side effects such as gastric bleeding, immunosuppression, osteoporosis and heart failure do not make them ideal agents for long-term use. Maintaining the AA/EPA blood ratio < 2 using omega-3 supplementation could provide an excellent therapeutic regime for reducing inflammation in the retina and the optic nerve, without any known side effects to date. Further studies are ongoing in order to get a better optimisation of the anti-inflammatory effect of omega-3 PUFAs and have an improved understanding on their mechanism of action and their long-term effects.

References

[1] Friedman DS, O'Colmain BJ, Munoz B, Tomany SC, McCarty C, de Jong PT, et al., Prevalence of age-related macular degeneration in the United States. *Archives of ophthalmology*. 2004;122(4):564-72.

[2] Wong WL, Su X, Li X, Cheung CM, Klein R, Cheng CY, et al., Global prevalence of age-related macular degeneration and disease burden projection for 2020 and 2040: a systematic review and meta-analysis. *The Lancet Global health*. 2014;2(2):e106-16.

[3] Resnikoff S, Pascolini D, Etya'ale D, Kocur I, Pararajasegaram R, Pokharel GP, et al., Global data on visual impairment in the year 2002. *Bulletin of the World Health Organization*. 2004;82(11):844-51.

[4] Miller JW. Age-related macular degeneration revisited--piecing the puzzle: the LXIX Edward Jackson memorial lecture. *American journal of ophthalmology*. 2013;155(1):1-35 e13.

[5] SanGiovanni JP, Agron E, Clemons TE, Chew EY. Omega-3 long-chain polyunsaturated fatty acid intake inversely associated with 12-year progression to advanced age-related macular degeneration. *Archives of ophthalmology*. 2009;127(1):110-2.

[6] Nag TC, Wadhwa S. Ultrastructure of the human retina in aging and various pathological states. *Micron*. 2012;43(7):759-81.

[7] Harman D. Aging: a theory based on free radical and radiation chemistry. *Journal of gerontology*. 1956;11(3):298-300.

[8] Nussenblatt RB, Lee RW, Chew E, Wei L, Liu B, Sen HN, et al., Immune responses in age-related macular degeneration and a possible long-term therapeutic strategy for prevention. *American journal of ophthalmology*. 2014;158(1):5-11 e2.

[9] Macaulay R, Akbar AN, Henson SM. The role of the T cell in age-related inflammation. *Age*. 2013;35(3):563-72.

[10] Ferrara DC, Merriam JE, Freund KB, Spaide RF, Takahashi BS, Zhitomirsky I, et al., Analysis of major alleles associated with age-

related macular degeneration in patients with multifocal choroiditis: strong association with complement factor H. *Archives of ophthalmology*. 2008;126(11):1562-6.

[11] Anderson DH, Radeke MJ, Gallo NB, Chapin EA, Johnson PT, Curletti CR, et al., The pivotal role of the complement system in aging and age-related macular degeneration: hypothesis re-visited. *Progress in retinal and eye research*. 2010;29(2):95-112.

[12] Klein RJ, Zeiss C, Chew EY, Tsai JY, Sackler RS, Haynes C, et al., Complement factor H polymorphism in age-related macular degeneration. *Science*. 2005;308(5720):385-9.

[13] Edwards AO, Ritter R, 3rd, Abel KJ, Manning A, Panhuysen C, Farrer LA. Complement factor H polymorphism and age-related macular degeneration. *Science*. 2005;308(5720):421-4.

[14] Gottfredsdottir MS, Sverrisson T, Musch DC, Stefansson E. Age related macular degeneration in monozygotic twins and their spouses in Iceland. *Acta ophthalmologica Scandinavica*. 1999;77(4):422-5.

[15] Xu H, Chen M, Forrester JV. Para-inflammation in the aging retina. *Progress in retinal and eye research*. 2009;28(5):348-68.

[16] Barouch FC, Miller JW. The role of inflammation and infection in age-related macular degeneration. *International ophthalmology clinics*. 2007;47(2):185-97.

[17] Penfold PL, Killingsworth MC, Sarks SH. Senile macular degeneration: the involvement of immunocompetent cells. *Graefe's archive for clinical and experimental ophthalmology = Albrecht von Graefes Archiv fur klinische und experimentelle Ophthalmologie*. 1985;223(2):69-76.

[18] Penfold PL, Liew SC, Madigan MC, Provis JM. Modulation of major histocompatibility complex class II expression in retinas with age-related macular degeneration. *Investigative ophthalmology & visual science*. 1997;38(10):2125-33.

[19] Seddon JM, Gensler G, Milton RC, Klein ML, Rifai N. Association between C-reactive protein and age-related macular degeneration. *JAMA : the journal of the American Medical Association*. 2004;291(6):704-10.

[20] Seddon JM, George S, Rosner B, Rifai N. Progression of age-related macular degeneration: prospective assessment of C-reactive protein, interleukin 6, and other cardiovascular biomarkers. *Archives of ophthalmology*. 2005;123(6):774-82.

[21] Klein R, Klein BE, Marino EK, Kuller LH, Furberg C, Burke GL, et al., Early age-related maculopathy in the cardiovascular health study. *Ophthalmology*. 2003;110(1):25-33.

[22] Klein R, Klein BE, Knudtson MD, Wong TY, Shankar A, Tsai MY. Systemic markers of inflammation, endothelial dysfunction, and age-related maculopathy. *American journal of ophthalmology.* 2005;140(1):35-44.

[23] Libby P, Ridker PM. Inflammation and atherosclerosis: role of C-reactive protein in risk assessment. *The American journal of medicine.* 2004;116 Suppl 6A:9S-16S.

[24] Cousins SW, Espinosa-Heidmann DG, Csaky KG. Monocyte activation in patients with age-related macular degeneration: a biomarker of risk for choroidal neovascularization? *Archives of ophthalmology.* 2004;122(7):1013-8.

[25] Gupta N, Brown KE, Milam AH. Activated microglia in human retinitis pigmentosa, late-onset retinal degeneration, and age-related macular degeneration. *Experimental eye research.* 2003;76(4):463-71.

[26] Chen M, Forrester JV, Xu H. Synthesis of complement factor H by retinal pigment epithelial cells is down-regulated by oxidized photoreceptor outer segments. *Experimental eye research.* 2007;84(4):635-45.

[27] Chen M, Muckersie E, Robertson M, Forrester JV, Xu H. Up-regulation of complement factor B in retinal pigment epithelial cells is accompanied by complement activation in the aged retina. *Experimental eye research.* 2008;87(6):543-50.

[28] Coffey PJ, Gias C, McDermott CJ, Lundh P, Pickering MC, Sethi C, et al., Complement factor H deficiency in aged mice causes retinal abnormalities and visual dysfunction. *Proceedings of the National Academy of Sciences of the United States of America.* 2007;104(42):16651-6.

[29] Sohn JH, Kaplan HJ, Suk HJ, Bora PS, Bora NS. Chronic low level complement activation within the eye is controlled by intraocular complement regulatory proteins. *Investigative ophthalmology & visual science.* 2000;41(11):3492-502.

[30] Sears B. The Anti-Inflammation Zone. New York: HarperCollins; 2005.

[31] Serhan CN, Arita M, Hong S, Gotlinger K. Resolvins, docosatrienes, and neuroprotectins, novel omega-3-derived mediators, and their endogenous aspirin-triggered epimers. *Lipids.* 2004;39(11):1125-32.

[32] Bazan NG. Omega-3 fatty acids, pro-inflammatory signaling and neuroprotection. *Current opinion in clinical nutrition and metabolic care.* 2007;10(2):136-41.

[33] Birch EE, Birch DG, Hoffman DR, Uauy R. Dietary essential fatty acid supply and visual acuity development. *Investigative ophthalmology & visual science*. 1992;33(11):3242-53.

[34] Gragoudas ES, Adamis AP, Cunningham ET, Jr., Feinsod M, Guyer DR, Group VISiONCT. Pegaptanib for neovascular age-related macular degeneration. *The New England journal of medicine*. 2004; 351(27):2805-16.

[35] Rofagha S, Bhisitkul RB, Boyer DS, Sadda SR, Zhang K, Group S-US. Seven-year outcomes in ranibizumab-treated patients in ANCHOR, MARINA, and HORIZON: a multicenter cohort study (SEVEN-UP). *Ophthalmology*. 2013;120(11):2292-9.

[36] Age-Related Eye Disease Study Research G. A randomized, placebo-controlled, clinical trial of high-dose supplementation with vitamins C and E and beta carotene for age-related cataract and vision loss: AREDS report no. 9. *Archives of ophthalmology*. 2001;119(10):1439-52.

[37] Buschini E, Fea AM, Lavia CA, Nassisi M, Pignata G, Zola M, et al., Recent developments in the management of dry age-related macular degeneration. *Clinical ophthalmology*. 2015;9:563-74.

[38] Mishra A, Chaudhary A, Sethi S. Oxidized omega-3 fatty acids inhibit NF-kappaB activation via a PPARalpha-dependent pathway. *Arteriosclerosis, thrombosis, and vascular biology*. 2004;24(9):1621-7.

[39] De Caterina R, Liao JK, Libby P. Fatty acid modulation of endothelial activation. *The American journal of clinical nutrition*. 2000;71(1 Suppl):213S-23S.

[40] Chen W, Esselman WJ, Jump DB, Busik JV. Anti-inflammatory effect of docosahexaenoic acid on cytokine-induced adhesion molecule expression in human retinal vascular endothelial cells. *Investigative ophthalmology & visual science*. 2005;46(11):4342-7.

[41] Rotstein NP, Politi LE, German OL, Girotti R. Protective effect of docosahexaenoic acid on oxidative stress-induced apoptosis of retina photoreceptors. *Investigative ophthalmology & visual science*. 2003;44(5):2252-9.

[42] Tuo J, Ross RJ, Herzlich AA, Shen D, Ding X, Zhou M, et al., A high omega-3 fatty acid diet reduces retinal lesions in a murine model of macular degeneration. *The American journal of pathology*. 2009;175(2):799-807.

[43] Calder PC. Long-chain n-3 fatty acids and inflammation: potential application in surgical and trauma patients. *Brazilian journal of medical and biological research = Revista brasileira de pesquisas medicas e*

biologicas / Sociedade Brasileira de Biofisica [et al]. 2003;36(4):433-46.

[44] Kapoor M, Kojima F, Yang L, Crofford LJ. Sequential induction of pro- and anti-inflammatory prostaglandins and peroxisome proliferators-activated receptor-gamma during normal wound healing: a time course study. *Prostaglandins, leukotrienes, and essential fatty acids*. 2007; 76(2):103-12.

[45] Ramkumar HL, Tuo J, Shen de F, Zhang J, Cao X, Chew EY, et al., Nutrient supplementation with n3 polyunsaturated fatty acids, lutein, and zeaxanthin decrease A2E accumulation and VEGF expression in the retinas of Ccl2/Cx3cr1-deficient mice on Crb1rd8 background. *The Journal of nutrition*. 2013;143(7):1129-35.

[46] Yanai R, Mulki L, Hasegawa E, Takeuchi K, Sweigard H, Suzuki J, et al., Cytochrome P450-generated metabolites derived from omega-3 fatty acids attenuate neovascularization. *Proceedings of the National Academy of Sciences of the United States of America*. 2014; 111 (26):9603-8.

[47] Webler AC, Michaelis UR, Popp R, Barbosa-Sicard E, Murugan A, Falck JR, et al., Epoxyeicosatrienoic acids are part of the VEGF-activated signaling cascade leading to angiogenesis. *American journal of physiology Cell physiology*. 2008;295(5):C1292-301.

[48] Zhang G, Panigrahy D, Mahakian LM, Yang J, Liu JY, Stephen Lee KS, et al., Epoxy metabolites of docosahexaenoic acid (DHA) inhibit angiogenesis, tumor growth, and metastasis. *Proceedings of the National Academy of Sciences of the United States of America*. 2013; 110(16):6530-5.

[49] Connor KM, SanGiovanni JP, Lofqvist C, Aderman CM, Chen J, Higuchi A, et al., Increased dietary intake of omega-3-polyunsaturated fatty acids reduces pathological retinal angiogenesis. *Nature medicine*. 2007;13(7):868-73.

[50] Rezende FA, Lapalme E, Qian CX, Smith LE, SanGiovanni JP, Sapieha P. Omega-3 supplementation combined with anti-vascular endothelial growth factor lowers vitreal levels of vascular endothelial growth factor in wet age-related macular degeneration. *American journal of ophthalmology*. 2014;158(5):1071-78.

[51] Tassos Georgiou, Anastasia Neokleous, Despina Nicolaou, Barry Sears. Pilot study for treating dry age-related macular degeneration (AMD) with high-dose omega-3 fatty acids. *PharmaNutrition*. 2014;2:8-11.

In: Age-Related Macular Degeneration
Editor: Melanie Carter
ISBN: 978-1-63483-329-5

Chapter IV

Bibliography

Age-related macular degeneration LCCN: 2012938196: Age-related macular degeneration / Frank G. Holz ... [et al.], editors. Edition: 2nd ed. Published/Created: Heidelberg; New York: Springer, c2013. Description: xx, 320 p.: ill. (some col.); 27 cm. ISBN: 9783642221064 (alk. paper) 3642221068 (alk. paper) 9783642221071 (ebook) 3642221076 (ebook) LC classification: RE661 .D3 A3222 2013 Related names: Holz, Frank G. Subjects: Retinal degeneration. Eye--Aging. Macular Degeneration. Notes: Includes bibliographical references and index. Dewey class no.: 617.7/35.

Age-related macular degeneration LCCN: 2014000607: Age-related macular degeneration (Alfaro) Main title: Age-related macular degeneration / editors, D. Virgil Alfaro, III, MD, Attending Surgeon, Retina Consultants of Charleston, South Carolina, Eric P. Jablon, MD, Partner, Retina Consultants of Charleston, Charleston, South Carolina, John B. Kerrison, MD, Partner, Retina Consultants of Charleston, Assistant Clinical Professor, Department of Ophthalmology, Medical University of South Carolina, Charleston, South Carolina, Kenneth A. Sharpe, MD, Partner, Retina Consultants of Charleston, Charleston, South Carolina, Monica Rodriguez-Fontal, MD, Clinical Trial Director, Retina Consultants of Charleston, Charleston, South Carolina. Edition: Second edition. Published/Produced: Philadelphia: Wolters Kluwer, [2015] Description: xii, 320 pages: illustrations (some color); 29 cm ISBN: 9781451151695 (hbk.) 1451151691 (hbk.) LC classification: RE661.D3 A32226 2015 Related names: Alfaro, D. Virgil, III, editor. Jablon, Eric P., editor. Kerrison, John B., editor. Sharpe, Kenneth A., editor. Rodriguez-Fontal, Monica, editor. Subjects: Macular Degeneration. Choroidal Neovascularization. Notes: Includes bibliographical references and index. Dewey class no.: 618.97/7735.

Age-related macular degeneration: current management LCCN: 2014018722: Age-related macular degeneration (Duker) Main title: Age-related macular degeneration: current management / edited by Jay S. Duker, Andre J. Witkin. Published/Produced: Thorofare, NJ: SLACK Incorporated, [2015] Description: xiv, 155 pages; 23 cm ISBN: 9781617116421 (pbk: alk. paper) LC classification: RE661.D3 A322263 2015 Related names: Duker, Jay S., 1958- editor. Witkin, Andre J., 1980- editor. Contents: Epidemiology, genetics & modifiable risk factors / Anita Agarwal -- Pathology of AMD / Nora M. Laver -- Clinical diagnosis, classification, and patient counseling / Sana Nadeem and Nadia K. Waheed -- Fundus photography and autofluorescence in age-related macular degeneration / S.K. Steven Houston III and Sunir J. Garg -- Mimickers of age-related macular degeneration / Kapil G. Kapoor and Sophie J. Bakri -- Intraviteral antivascular endothelial growth factor therapy / Mariana R. Thorell and Philip J. Rosenfeld -- Thermal laser and photodynamic therapy / Llewelyn J. Rao and Lawrence J. Singerman -- Treatment failures: nonresponders, tolerance, and tachyphylaxis / Christopher j. Brady and Chirag P. Shah -- Surgery for age-related macular degeneration / Christopher J. Brady, MD and Carl D. Regillo -- Future therapies / Roger A. Goldberg and Jeffrey S. Heier. Subjects: Macular Degeneration. Age Factors. Notes: Includes bibliographical references and index. Dewey class no.: 617.7/35.

Angiotensin: new research LCCN: 2011038153: Angiotensin: new research / Sota Harada and Itsuki Moi, editors. Published/Created: New York: Nova Science

Publishers c2012. Description: xiii, 336 p.: ill. (some col.); 27 cm. ISBN: 9781621007739 (hardcover) 1621007731 (hardcover) LC classification: QP572.A54 A54 2012 Related names: Harada, Sota. Moi, Itsuki. Contents: The influence of renin-angiotensin-aldosterone system (Raas) genes on hypertrophy development in hypertrophic cardiomyopathy / Nadia Carstens ... [et al.] -- Cardiac renin angiotensin system and exercise training / G. Barauna, T. Fernandes, E.M. Oliveira -- The role of the counter-regulatory renin angiotensin system in the heart / Stuart A. Nicklin, Monica Flores-Muñoz -- New roles for the neuropeptide angiotensin II: stress and drug abuse / Bregonzio Claudia ... [et al.] -- Angiotensin I-converting enzyme inhibitory activity in lima bean (Phaseolus lunatus) protein hydrolysates produced with alcalase or pepsin-pancreatin / L. Chel-Guerrero ... [et al.] -- Pathophysiological role of angiotensin-converting enzyme in the cardiovascular system / Maira Segura-Campos ... [et al.] -- Redox regulation of angiotensin receptor signaling in the heart / Motohiro Nishida ... [et al.] -- Cross talk between angiotensin II and Ppar- during cardiac remodeling / Laxmansa C. Katwa ... [et al.] -- Cell swelling activates ionic channels and decreases the gap junction permeability: implications for metabolic cooperation between cells and for cardiac arrhythmias / Walmor C. De Mello -- Diabetic cardiomyopathy: role of renin angiotensin system / Kenichi Watanabe ... [et al.] -- Hypertension a risk factor for age-related macular degeneration: role of angiotensin II / Maria E. Marin-Castaño -- Do angiotensin II receptors play a role in cerebellar development and neuronal migration? / Susana I. Sánchez ... [et al.] -- Advances in non-invasive imaging of renin angiotensin system / Kumiko C. Mackasey, Rob S. Beanlands and Jean N. DaSilva. Subjects: Angiotensins. Notes: Includes bibliographical references and index. Series: Protein science and engineering Endocrinology research and clinical developments Dewey class no.: 572/.65.

Carotenoids and vitamin A in translational medicine LCCN: 2012038319 Carotenoids and vitamin A in translational medicine / edited by Olaf Sommerburg, Werner Siems, and Klaus Kraemer. Published/Created: Boca Raton,

FL: CRC Press/Taylor & Francis Group, c2013. Description: xxxii, 404 p.: ill.; 24 cm. ISBN: 9781439855263 (hardcover: alk. paper) LC classification: QP671.C3 C376 2013 Related names: Sommerburg, Olaf. Siems, Werner. Kraemer, Klaus, 1960- Contents: Hepatic retinoid metabolism: what is known and what is not / Jason J. Yuen, Jian Zhang Yang and William S. Blaner -- Metabolism of vitamin A in white adipose tissue and obesity / Shanmugam M. Jeyakumar ... [et al.] -- Retinoids in the treatment of psoriasis / Neil Borkar, Alan Menter -- Age-related macular degeneration and carotenoids / Nilesh Kalariya, Werner G. Siems, Frederik van Kuijk -- Lutein, zeaxanthin and vision across the lifespan / Billy R. Hammond, Jr. -- Assuring vitamin A adequacy to prevent eye and ear disorders / Tripper Sauer, Susan Emmett, Keith P. West, Jr. -- Disturbed accumulation and abnormal distribution of macular pigment in retinal disorders / Tos TJM Berendschot, Peter Charbel Issa, Thomas Theelen -- Retinol and otitis media / Akeem Olawale Lasisi -- Carotenoids and mutagenesis / P. M. Eckl ... [et al.] -- Carotenoids and vitamin A in lung cancer prevention / Anita Ratnasari Iskandar, Xiang-Dong Wang -- Is the effect of [Beta]-carotene on prostate cancer cells dependent on their androgen sensitivity? / Joanna Dulinska-Litewka ... [et al.] -- Carotenoid intake and supplementation in cancer: pro and con / W.G. Siems ... [et al.] -- Vitamin A in lung development and function / Hans-Konrad Biesalski -- Vitamin A and carotenoids in lung diseases / Olaf Sommerburg, Werner Siems -- Chemopreventive effects of astaxanthin on inflammatory bowel disease and inflammation-related colon carcinogenesis / Masashi Hosokawa, Yumiko Yasui -- Carotenoids in laboratory medicine / Ingolf Schimke, Michael Vogeser -- Supply of vitamin A in developing countries / Klaus Kraemer -- Vitamin A intakes are below recommendations in a significant proportion of the population in the western world / Hans-Konrad Biesalski, Barbara Trösch, Petra Weber -- Critical appraisal of vitamin A supplementation program in India / Dechenla Tshering Bhutia, Saskia De Pee -- Consequences of common genetic variations on SS-carotene cleavage for vitamin A supply / Georg Lietz, Anthony

Oxley, Christine Boesch-Saadatmandi. Subjects: Carotenoids--Therapeutic use. Vitamin A--Physiological effect. Vitamin A in the body. Vitamin A--therapeutic use. Carotenoids--therapeutic use. Translational Medical Research. Vitamin A--supply & distribution. Vitamin A Deficiency--drug therapy. Notes: Includes bibliographical references and index. Series: Oxidative stress and disease; 33 Oxidative stress and disease; 33. Dewey class no.: 615.3/28.

Cell-based therapy for retinal degenerative disease LCCN: 2014004771: Cell-based therapy for retinal degenerative disease / volume editors, Ricardo P. Casaroli-Marano, Barcelona, Marco A. Zarbin, Newark, NJ. Published/Produced: Basel; New York: Karger, [2014] ©2014 Description: xii, 205 pages: illustrations (chiefly color); 26 cm. ISBN: 9783318025842 (hard cover: alk. paper) 3318025844 (hard cover: alk. paper) LC classification: RE661.D3 C45 2014 Related names: Casaroli-Marano, Ricardo Pedro, editor of compilation. Zarbin, Marco A., editor of compilation. Contents: Age-related macular degeneration: clinical findings, histopathology, and imaging techniques / Zarbin, M.A.; Casaroli-Marano, R.; Rosenfeld, P.J. -- General pathophysiology in retinal degeneration / Wert, K.J.; Lin, J.H.; Tsang, S.H. -- Juvenile-onset macular degeneration and allied disorders / North, V.; Gelman, R.; Tsang, S.H. -- Diagnosis and complementary examinations / Menghini, M.; Duncan, J.L. -- Stem cells for retinal repair / Stern, J.; Temple, S. -- Differentiation of pluripotent stem cells into retinal pigmented epithelium / Croze, R.H.; Clegg, D.O. -- Neural retinal regeneration with pluripotent stem cells / Ramsden, C.M.; Powner, M.B.; Carr, A.-J.F.; Smart, M.J.K.; da Cruz, L.; Coffey, P.J. -- Stem cells, retinal ganglion cells and glaucoma / Sluch, V.M.; Zack, D.J. -- Stem cells: immunology and immunomodulation / Tena, A.; Sachs, D.H. -- Biochemical restoration of aged human Bruch's membrane: experimental studies to improve retinal pigment epithelium transplant survival and differentiation / Sugino, I.K.; Sun, Q.; Cheewatrakoolpong, N.; Malcuit, C.; Zarbin, M.A. -- Approaches to cell delivery: substrates and scaffolds for cell therapy / Kundu, J.; Michaelson, A.; Baranov, P.; Young, M.J.;

Carrier, R.L. -- Microdevice-based cell therapy for age-related macular degeneration / Lu, B.; Tai, Y.-C.; Humayun, M.S. -- Cell and gene therapy / Rao, R.C.; Zacks, D.N. -- Cellular manufacturing for clinical applications / Sheu, J.; Klassen, H.; Bauer, G. -- Regulatory issues in cell-based therapy for clinical purposes / Casaroli-Marano, R.P.; Tabera, J.; Vilarrodona, A.; Trias, E. Subjects: Retinal degeneration--Treatment. Cellular therapy. Stem cells--Therapeutic use. Retinal Degeneration--therapy. Stem Cell Transplantation. Notes: Includes bibliographical references and index. Series: Developments in ophthalmology, 0250-3751; vol. 53 Developments in ophthalmology; v. 53. 0250-3751 Dewey class no.: 617.7/35.

Epidemiology of eye disease LCCN: 2012405447: Epidemiology of eye disease / edited by Gordon J. Johnson ... [et al.]. Edition: 3rd ed. Published/Created: London: Imperial College Press; Singapore; Hackensack, NJ: Distributed by World Scientific, 2012. Description: xix, 645 p., [10] p. of plates: ill. (some col.), maps (some col.); 28 cm. ISBN: 9781848166257 1848166257 LC classification: RE46 .E86 2012 Related names: Johnson, Gordon J. Contents: SECTION 1: INTRODUCTION -- 1. Prevalence, incidence and distribution of visual impairment / G. V. S. Murthy and Gordon Johnson -- SECTION 2: METHODOLOGY -- 2. Epidemiological research methods: an outline / Darwin Minassian and Hannah Kuper -- 3. Cross-sectional studies / Darwin Minassian and Hannah Kuper -- 4. Case-control studies / Astrid Fletcher -- 5. Cohort studies / Emily Herrett and Hannah Kuper -- 6. Genetic epidemiology / Christopher Hammond -- 7. Clinical trials / Maureen Maguire -- 8. Screening in ophthalmology / Richarad Wormald and Robert Lindfield -- 9. Research synthesis / Jennifer Evans and Richard Wormald -- SECTION 3: SPECIFIC ENTITIES -- 10. Age-related cataract / Emily Gower and Sheila West -- 11. Epidemiology of refractive errors and presbyopia / Leslie Hyman and Ilesh Patel -- 12. The epidemiology of low vision / Gary Rubin -- 13. Glaucoma / Paul Foster and Harry Quigley -- 14. Visual impairment and blindness in children -- 14a.Magnitude and causes / Clare Gilbert and Jugnoo Rahi -- 14b.Vitamin A deficiency disorders / Keith West Jr. and Alfred Sommer -- 14c.Ophthalmia neonatorum / Volker Klauss and Ulrich Schaller -- Contents note continued: 14d.Corneal disease in children / Clare Gilbert -- 14e.Cataract in children / Edward Wilson and Rupal Trivedi -- 14f.Glaucoma in children / Maria Papadopoulos and Peng Tee Khaw -- 14g.Retinopathy of prematurity / Subhadra Jalali and Graham Quinn -- 15.Dry eye disease / Debra Schaumberg and Giulio Ferrari -- 16.Corneal and external diseases -- 16a.Microbial keratitis / John Whitcher, Mathua Srinivasan and Madan Upadhyay -- 16b.Viral infectious keratoconjunctivitis / Stephen Tuft and Bita Manzouri -- 16c.Acanthamoeba keratitis / Stephanie Watson and John Dart -- 16d.Ocular manifestations of leprosy / Paul Courtright and Susan Lewallen -- 16e.Vernal keratoconjunctivitis / Stephen Tuft and Bita Manzouri -- 16f.Mooren's ulcer / Stephen Tuft and Bita Manzouri -- 16g.Climatic droplet keratopathy / Gordon Johnson -- 16h.Pterygium / Gordon Johnson -- 17.Epidemiology of trachoma / Sheila West and Robin Bailey -- 18.Onchocerciasis / Adrian Hopkins -- 19.The epidemiology of uveitis / Jennifer Thorne and Douglas Jabs -- 20.Ocular complications of human immunodeficiency virus infection / John Kempen -- 21.Diabetic retinopathy / Barbara Klein and Ronald Klein -- 22.Age-related macular degeneration / Jennifer Evans and Tien Wong -- Contents note continued: SECTION 4 PREVENTION STRATEGIES -- 23.From epidemiology to programme / Robert Lindfield, Hans Limburg and Allen Foster -- 24.Global initiative for the elimination of avoidable blindness / Robert Lindfield, Ivo Kocur, Hans Limburg and Allen Foster. Subjects: Eye--Diseases--Epidemology. Blindness--Prevention and control. Notes: Previous ed.: 2003. Includes biliographical references and index. Dewey class no.: 617.7.

Epistasis: methods and protocols
LCCN: 2014953909: Epistasis: methods and protocols / edited by Jason H. Moore, Department of Genetics, Institute of Quantitative Biomedical Sciences, Geisel School of Medicine, Hanover, NH, USA, Scott M. Williams, Department of Genetics, Institute of Quantitative Biomedical

Sciences, Geisel School of Medicine, Hanover, NH, USA. Published/Produced: New York: Humana Press, [2015] ©2015 Description: x, 350 pages: illustrations (some color); 27 cm Links: Table of contents http: //www.loc.gov/catdir/enhancements/fy1502/2014953909-t.html ISBN: 9781493921546 (alk. paper) 1493921541 (alk. paper) LC classification: QH450.7 .E657 2015 QH506 .M45 v.1253 Related names: Moore, Jason H., editor. Williams, Scott M. (Scott Matthew), editor. Springer Science+Business Media, copyright holder. Contents: Long-term selection experiments: epistasis and the response to selection / Charles Goodnight -- Finding the epistasis needles in the genome-wide haystack / Marylyn D. Ritchie -- Biological knowledge-driven analysis of epistasis in human GWAS with application to lipid traits / Li Ma, Alon Keinan, and Andrew G. Clark -- Epistasis for quantitative traits in Drosophila / Trudy F. C. Mackay -- Epistasis in the risk of human neuropsychiatric disease / Scott M. Williams -- On the partitioning of genetic variance with epistasis / José M. Álvarez-Castro and Arnaud Le Rouzic -- Measuring gene interactions / Thomas F. Hansen -- Two rules for the detection and quantification of epistasis and other interaction effects / Günter P. Wagner -- Direct approach to modeling epistasis / Rong-Cai Yang -- Capacitating epistasis - detection and role in the genetic architecture of complex traits / Mats E. Pettersson and Örjan Carlborg -- Compositional epistasis: an epidemiologic perspective / Etsuji Suzuki and Tyler J. VanderWeele -- Identification of genome-wide SNP-SNP and SNP-clinical boolean interactions in age-related macular degeneration / Carlos Riveros [and eight others] -- Epistasis analysis using information theory / Jason H. Moore and Ting Hu -- Genome-wide epistasis and pleiotropy characterized by the bipartite human phenotype network / Christian Darabos and Jason H. Moore -- Network theory for data-driven epistasis networks / Caleb A. Lareau and Brett A. McKinney -- Epistasis analysis using multifactor dimensionality reduction / Jason H. Moore and Peter C. Andrews -- Epistasis analysis using ReliefF / Jason H. Moore -- Epistasis analysis using artificial intelligence / Jason H. Moore and Doug P. Hill. Subjects: Epistasis (Genetics)--Laboratory manuals.

Epistasis, Genetic--Laboratory Manuals. Epistasis (Genetics) Form/Genre: Laboratory Manuals. Handbooks, manuals, etc. Notes: Includes bibliographical references and index. Series: Methods in Molecular Biology, 1064-3745; 1253 Springer protocols, 1949-2448 Methods in molecular biology (Clifton, N.J.); v. 1253, 1064-3745 Springer protocols (Series), 1949-2448.

Genetic diseases of the eye LCCN: 2011010629: Genetic diseases of the eye / edited by Elias I. Traboulsi. Edition: 2nd ed. Published/Created: New York: Oxford University Press, c2012. Description: xv, 923 p.: ill. (some col.); 28 cm. ISBN: 9780195326147 (alk. paper) 0195326148 (alk. paper) LC classification: RE906 .G39 2012 Related names: Traboulsi, Elias I. Contents: Section One: Malformations -- Chapter 1: Embryology of the eye and the role of developmental genes / Olof H. Sundin -- Chapter 2: Teratogens and ocular malformations / Kerstin Strömland and Marilyn T. Miller -- Chapter 3: Malformations of the ocular adnexae / Craig Lewis, Katrinka L. Heher, James A. Katowitz, and Elias I. Traboulsi -- Chapter 4: Nanophthalmos / Eduardo Duarte Silva and Olof H. Sundin -- Chapter 5: Anophthalmia, colobomatous, microphthalmia, and optic fissure closure defects / Brian P. Brooks and Elias I. Traboulsi -- Chapter 6: Cornea plana / Arif O. Khan -- Chapter 7: Malformations of the anterior segment of the eye / James E. H. Smith and Elias I. Traboulsi -- Chapter 8: Aniridia / Reecha Sachdeva and Elias I. Traboulsi -- Chapter 9: Congenital anomalies of the optic nerve / Brian P. Brooks and Elias I. Traboulsi -- Chapter 10: Congenital abnormalities of the retinal pigment epithelium / Arturo Santos and Elias I. Traboulsi -- Chapter 11: Prenatal imaging of the eye and ocular adnexae / Erin Broaddus, Donna Patno, Janet Reid, Jeffrey Chapa, Elias I. Traboulsi, and Arun D. Singh -- Chapter 12: Ocular manifestations of syndromes with craniofacial abnormalities / Wadih M. Zein, Amy Feldman Lewanda, Elias I. Traboulsi, and Ethylin Wang Jabs -- Chapter 13: Ocular manifestations of chromosomal abnormalities / Sorath Noorani Siddiqui, Alex V. Levin, Matt Rusinek, Joanne E. Sutherland, and Anthony G. Quinn -- Section Two: Refractive errors, cornea, glaucoma, and cataracts

-- Chapter 14: Inheritance of refractive errors / Wadih M. Zein and Arlene V. Drack -- Chapter 15: Corneal dystrophies / Walter Lisch, Elias I. Traboulsi, and Dimitri T. Azar -- Chapter 16: The genetics of keratoconus / Marzena Gajecka -- Chapter 17: Molecular genetics of primary congenital glaucoma / Roshanak Sharafieh, Anne H. Child, and Mansoor Sarfarazi -- Chapter 18: Molecular genetics of primary open-angle glaucoma / Anne H. Child, Filipe M. Pereira da Silva, Jose Aragon-Martin, Roshanak Sharafieh, and Mansoor Sarfarazi -- Chapter 19: Genetics of congenital cataracts / Madhavan Jagadeesan and Elise Héon -- Section Three: Retina and optic nerve -- Chapter 20: Retinal function testing and genetic disease / Luis A.R. Gabriel, Neal S. Peachey, and Janet S. Sunness -- Chapter 21: Genetic pathways in retinal degenerations and targets for therapy / Loh-Shan Bryan Leung, Vinod Babu Voleti, Jonathan H. Lin, and Stephen H. Tsang -- Chapter 22: Proteomic biomarkers for age-related macular degeneration / John W. Crabb -- Chapter 23: Retinitis pigmentosa / Henry A. Ferreyra and John R. Heckenlively -- Chapter 24: Juvenile retinoschisis / Paul A. Sieving and Lucia Ziccardi -- Chapter 25: Achromatopsia-rod monochromacy / Susanne Kohl -- Chapter 26: Cone dysfunction syndromes, cone dystrophies, and cone-rod degenerations / Elias I. Traboulsi -- Chapter 27: North Carolina macular dystrophy/MCRD1 / Kean T. Oh and Kent Small -- Chapter 28: Bestrophinopathies / Bart P. Leroy -- Chapter 29: NR2E3-linked retinal degenerations: enhanced S-cone sensitivity syndrome, Goldmann-Favre syndrome, clumped pigmentary retinal degeneration, and retinitis pigmentosa / Daniel F. Schorderet, Neena Haider, and Pascal Escher -- Chapter 30: Disorders of color vision / Samir S. Deeb and Arno G. Motulsky -- Chapter 31: Stargardt disease / Aimee V. Chappelow and Elias I. Traboulsi -- Chapter 32: Congenital stationary night blindness / Elias I. Traboulsi, Bart P. Leroy, and Christina Zeitz -- Chapter 33: Choroideremia / Ian M. MacDonald and Miguel C. Seabra -- Chapter 34: Leber congenital amaurosis: clinical, genetic, and therapeutic perspectives / Robert K. Koenekoop, Frans P.M. Cremers, Irma Lopez, and Anneke I. den Hollander -- Chapter 35: Familial exudative vitreoretinopathy, Norrie disease, and other developmental retinal vascular disorders / Johane M. Robitaille, Duane L. Guernsey, and Elias I. Traboulsi -- Chapter 36: Hereditary vitreoretinopathies / Daniel F. Rosberger, Ravi D. Patel, and Elias I. Traboulsi -- Chapter 37: Genetics of age-related maculopathy / Oluwatoyin Fafowora and Michael B. Gorin -- Chapter 38: Pattern dystrophies of the RPE / Kean T. Oh -- Chapter 39: Hereditary optic neuropathies / David A. Mackey -- Chapter 40: Pigmentary retinopathy in systemic inherited disease / Ying Qian, Richard Alan Lewis, and Elias I. Traboulsi -- Section Four: Eye movement disorders -- Chapter 41: The genetics of nystagmus and associated inherited diseases / Shery Thomas and Irene Gottlob -- Chapter 42: The genetics of strabismus and associated disorders / Gena Heidary, Elias I. Traboulsi, and Elizabeth C. Engle -- Section Five: Systemic disease and the eye -- Chapter 43: Ectopia lentis and associated systemic disease / Elias I. Traboulsi and Suneel S. Apte -- Chapter 44: Peroxisomal disorders / Mark E. Pennesi and Richard G. Weleber -- Chapter 45: Albinism / Reecha Sachdeva, Lisa S. Abrams, and Elias I. Traboulsi -- Chapter 46: The phakomatoses / Michelle M. Ariss, Nicola K. Ragge, Manikum Moodley, and Elias I. Traboulsi -- Section Six: Cancer genetics and the eye -- Chapter 47: Systemic associations of eyelid tumors / Michelle M. Ariss, Elias I. Traboulsi, and Arun D. Singh -- Chapter 48: Genetic aspects of uveal melanoma / Werner Wackernagel and Arun D. Singh -- Chapter 49: Genetics of retinoblastoma / Emily Edelman, Rubens N. Belfort, Evelyn X. Fu, and Arun D. Singh -- Section Seven: Treatment -- Chapter 50: Vision rehabilitation of the patient with genetic eye disease / Joseph L. DeRose -- Chapter 51: Genetic counseling for genetic eye disorders / Joanne E. Sutherland -- Chapter 52: Gene therapy for ocular diseases / Ben J. Kim and Nadia K. Waheed. Subjects: Eye--Diseases--Genetic aspects. Eye--Abnormalities. Eye Diseases, Hereditary. Notes: Includes bibliographical references and index. Series: Oxford monographs on medical genetics; 61 Oxford monographs on medical genetics; no. 61. Dewey class no.: 617.7/042

Handbook of clinical trials in ophthalmology LCCN: 2014395245: Handbook of clinical trials in ophthalmology / AK Gupta ... [et al.]. Published/Created: New Delhi; Philadelphia: Jaypee Brothers Medical Pub., c2014. Description: xx, 314 p.; 24 cm. ISBN: 9789350907740 9350907747 LC classification: RE58 .G87 2014 Related names: Aggarwal, Vinod Kumar. Goel, Neha. Contents: Clinical trials in cornea -- Clinical trials in glaucoma -- Clinical trials in diabetic retinopathy-I -- Clinical trials in diabetic retinopathy-II -- Clinical trials in diabetic retinopathy-III -- Clinical trials in diabetic retinopathy-IV -- Clinical trials in retinal vascular occlusions -- Clinical trials in age-related macular degeneration-I -- Clinical trials in age-related macular degeneration-II -- Clinical trials in age-related macular degeneration-III -- Clinical trials in age-related macular degeneration-IV -- Clinical trials in retinal detachment -- Endophthalmitis vitrectomy study (EVS) -- Clinical trials in retinopathy of prematurity (ROP) -- Infant aphakia treatment study (IATS) -- Clinical trials in amblyopia -- Clinical trials in neuro-ophthalmology -- Collaborative ocular melanoma study (COMS) -- Clinical trials in ocular complications of AIDS. Subjects: Eye--Diseases--Treatment. Clinical trials. Notes: Includes bibliographical references. Dewey class no.: 617.7/1.

Ocular drug delivery systems: barriers and application of nanoparticulate systems LCCN: 2012015794: Ocular drug delivery systems: barriers and application of nanoparticulate systems / editors, Deepak Thassu and Gerald J. Chader. Published/Created: Boca Raton: CRC Press/Taylor & Francis, 2013. Description: xv, 447 p.: ill.; 24 cm. ISBN: 9781439848005 (hardcover: alk. paper) LC classification: RE48 .O33 2013 Related names: Thassu, Deepak. Chader, Gerald. Contents: Nanoparticulate-based therapeutics: an overview / Deepak Thassu, Kris Holt, and Gerald J. Chader -- Eye anatomy, physiology and ocular barriers: basic considerations for drug delivery / Gerald J. Chader and Deepak Thassu -- Animal models to evaluate nanoparticulate drug delivery systems / Vinson M. Wang, Jingsheng Tuo and Chi-Chao Chan -- Computer modeling for ocular drug delivery / Paul J. Missel -- Challenges in the development of nanotechnology-based ophthalmic systems / Banu S. Zolnik, Sherizaq Baksh and Nakissa Sadrieh -- Blood-retinal barrier: the fundamentals / Rosa Fernandes, Andreia Goncalves and Jose Cunha-Vaz -- Barriers to transscleral drug delivery to the retina / Alexandra Almazan ... [et al.] -- Drug delivery to the vitreous humor / Francine Behar-Cohen -- Transscleral and suprachoroidal drug delivery / Damian E. Berezovsky and Henry Edelhauser -- Protein drug delivery to retina and choroid / Hogwen M. Rivers and Patrick M. Hughes -- Transscleral drug delivery to the posterior segment of the eye / Bernard F. Godley ... [et al.] -- Drug delivery to the suprachoroidal space / Sung Won Cho and Timothy Olsen -- Use of nanoparticles in the treatment of age-related macular degeneration, glaucoma and other degenerative retinal diseases / Marco A. Zarbin ... [et al.] -- Drug molecule characteristics, drug delivery strategies and their impact on ophthalmic drug delivery to anterior vs. posterior segments / Kay D. Rittenhouse -- Surface modulation of nanoparticles / Deepak Thassu and Kris Holt -- Ophthalmic drug inserts / Kay Rittenhouse and Deepak Thassu -- Nanoparticles for the treatment of retinal diseases / Amin Famili and Uday Kompella -- Gene-based medicine for ocular diseases -- Shannon M. Conley and Muna I. Naash -- Stealth-type polymeric nanoparticles with encapsulated betamethasone phosphate for treatment of intraocular inflammation / Tsutomu Sakai, Tsutomu Ishihara and Megumu Higaki -- Nanotechnology-guided imaging of retinal vascular disease / Joshua R. Trantum and Ashwath Jayagopal -- Photoresponsive polymers for ocular drug delivery / Laura A. Wells and Heather Sheardown -- Nanosuspensions in ocular drug delivery systems / Himanshu Bhattacharjee and Sonia Bedi. Subjects: Eye Diseases--drug therapy. Blood-Retinal Barrier. Drug Delivery Systems--methods. Nanoparticles--therapeutic use. Notes: Includes bibliographical references and index. Dewey class no.: 617.7/061.

Ophthalmic radiation therapy: techniques and applications LCCN: 2013024151: Ophthalmic radiation therapy: techniques and applications / volume editors, Arun D. Singh,

David E. Pelayes, Stefan Seregard, Roger Macklis. Published/Produced: Basel: Karger, 2013. Projected pub date: 1111 Description: p.; cm. ISBN: 9783318024401 (hard cover: alk. paper) LC classification: RC280.E9 Related names: Singh, Arun D., editor of compilation. Pelayes, David E., editor of compilation. Seregard, Stefan, editor of compilation. Macklis, Roger M., editor of compilation. Contents: Introduction to radiotherapy and standard teletherapy techniques / Balagamwala, E.H., Macklis, R., Singh, A.D. -- Teletherapy: advanced techniques / Stockham, A., Balagamwala, E.H., Singh, A.D., Macklis, R. -- Brachytherapy / Marwaha, G., Macklis, R., Singh, A.D., Wilkinson, A. -- Radiation therapy: uveal tumors / Seregard, S., Pelayes, D.E., Singh, A.D. -- Radiation therapy retinal tumors / Sethi, R.V., MacDonald, S.M., Kim, D.Y., Mukai, S. -- Radiation therapy: age-related macular degeneration / Medina, C.A., Ehlers, J.P. -- Radiation therapy: conjunctival and eyelid tumors / Aronow, M.E., Singh, A.D. -- Radiation therapy: orbital tumors / Marwaha, G., Macklis, R., Singh, A.D. -- Radiotherapy: anterior segment complications / Medina, C., Singh, A.D. -- Radiation therapy: posterior segment complications / Seregard, S., Pelayes, D.E., Singh, A.D. Subjects: Eye Neoplasms--radiotherapy. Radiotherapy--methods. Notes: Includes bibliographical references and indexes. Series: Developments in ophthalmology, 0250-3751; vol. 52 Developments in ophthalmology; v. 52. 0250-3751 Dewey class no.: 616.99/484.

Ophthalmology at a glance LCCN: 2013038050: Ophthalmology at a glance / Jane Olver, Lorraine Cassidy, Gurjeet Jutley, Laura Crawley. Edition: Second edition. Published/Produced: Chichester, West Sussex, UK: John Wiley & Sons Inc., 2014. Projected pub date: 1404 Description: p.; cm. ISBN: 9781405184731 (paper) LC classification: RE50 Related names: Cassidy, Lorraine, author. Crawley, Laura, active 2014, author. Jutley, Gurjeet, author. Summary: "This fully updated new edition teaches you exactly what they need to know, starting with taking a history and examination, before moving through specific conditions and their treatment. It includes new chapters on tropical ophthalmology, ocular oncology and giant cell arteritis, extra chapters on red eye and painful loss of vision and discussion of new scientific findings surrounding the cornea. Fully supported by a brand new companion website at ataglanceseries.com/ophthal featuring 20 clinical cases, images from the book as PowerPoint slides, and digital flashcards"--Provided by publisher. Contents: Machine generated contents note: Section 1 Principles of ophthalmology 1 Introduction: what is ophthalmology? 2 Medical student aims 3 Social and occupational aspects of vision Section 2 Ophthalmic history and examination 4 Taking the history and recording the findings 5 Visual acuity in adults 6 Examination of visual fields 7 Other visual functions Section 3 Correction of refractive errors 8 Basic optics and refraction 9 Glasses, contact lenses, and low-vision aids Section 4 Basic eye examination 10 External eye and anterior segment 11 Posterior segment and retina 12 Use of eye drops Section 5 Acute ophthalmology 13 Red eye: the conjunctiva 14 Red eye: the cornea 15 Red eye: the sclera and episclera 16 Ophthalmic trauma principles and management of chemical injuries 17 Specific features of blunt and sharp injuries 18 Sudden painful loss of vision in a non-inflamed eye 19 Loss of vision in the inflamed eye 20 Sudden painless loss of vision Section 6 Gradual loss of vision 21 Gradual loss of vision Section 7 Sub-specialty: Paediatric ophthalmology 22 Visual acuity in children 23 Strabismus (squints) 24 Neonates 25 Infants and older children Section 8 Sub-specialty: Eyelid, lacrimal and orbit 26 Common eyelid lumps 27 Common eyelid malpositions 28 Lacrimal (tearing) 29 Basic orbital assessment 30 Orbit disease, thyroid disease and facial palsy 31 Ocular oncology Section 9 Sub-specialty: External eye disease 32 Common conditions affecting the external eye 33 Common conditions affecting the cornea 34 Medical uses of contact lenses 35 Corneal and laser refractive surgery Section 10 Sub-specialty: Corneal, refractive and cataract surgery 36 Cataract assessment 37 Cataract surgery 38 Cataract surgery postoperative care Section 11 Sub-specialty: Glaucoma 39 Glaucoma: the basics 40 Detecting glaucoma 41 Medical and surgical treatment of glaucoma Section 12 Sub-specialty: Vitreo-retinal 42 Retinal detachment 43 Retinal and choroidial anatomy and imaging 44 Inherited retinal dystrophies and age-related macular degeneration 45 Diabetic retinopathy classification and typical lesions 46 Diabetic retinopathy treatment 47 Retinal vein occlusion 48 Retinal artery obstruction 49 HIV infection and AIDS 50 Tropical ophthalmology: tropical eye disease 51 Tropical ophthalmology: global eye health Section 13 Sub-specialty: Neuro-ophthalmology 52 Pupil abnormalities 53 Optic nerve disease 54 Cranial nerve palsies and eye movement disorders 55 Visual field defects 56 Giant cell arteritis Appendix I: Red eye. Subjects: Eye Diseases--Examination Questions. Ophthalmology--methods--Examination Questions. Notes: Includes bibliographical references and index. Additional formats: Online version: Olver, Jane, author. Ophthalmology at a glance Second edition. Chichester, West Sussex, UK: John Wiley & Sons Inc., 2014 9781118797235 (DLC) 2013043086 Series: At a glance At a glance series (Oxford, England) Dewey class no.: 617.7.

Optical coherence tomography: a clinical and technical update LCCN: 2012938448: Optical coherence tomography: a clinical and technical update / Rui Bernardes, José Cunha-Vaz, editors. Published/Created: Heidelberg; New York: Springer, 2012. Description: xv, 255 p.: ill. (some col.); 24 cm. ISBN: 9783642274091 (alk. paper) 3642274099 (alk. paper) 9783642274107 (ebk.) LC classification: RE79.I42 O68 2012 Related names: Bernardes, Rui. Cunha-Vaz, José. Summary: Optical coherence tomography represents the ultimate noninvasive ocular imaging technique, though it has been in the field for over two decades. This book encompasses medical and technical developments and recent achievements. Contributors cover the field of application from the anterior to the posterior ocular segments (Part I) and present a comprehensive review on the development of OCT. Important developments towards clinical applications are covered in Part II, ranging from adaptive optics to the integration on a slit-lamp, and passing through new structural and functional information extraction from OCT data. The book is intended to be informative, coherent and comprehensive for the medical and technical communities.-- Source other than Library of Congress. Contents: Diabetic macular edema / Conceição Lobo, Isabel Pires and José Cunha-Vaz -- Ischemia / Suk Ho Byeon, Min Kim and Oh Woong Kwon -- Optical coherence tomography and visual acuity: photoreceptor loss / Adzura Salam, Ute Ellen Kathrin Wolf-Schnurrbusch and Sebastian Wolf -- Optical coherence tomography updates on clinical and technical developments. Age-related macular degeneration: drusen and geographic atrophy / Monika Fleckenstein, Steffen Schmitz-Valckenberg and Frank G. Holz -- Optical coherence tomography in glaucoma / Fatmire Berisha, Esther M. Hoffmann and Norbert Pfeiffer -- Anterior segment OCT imaging / Alexandre Denoyer, Antoine Labbé and Christophe Baudouin -- Optical coherence tomography: a concept review / Pedro Serranho, António Miguel Morgado and Rui Bernardes -- Evaluation of the blood-retinal barrier with optical coherence tomography / Rui Bernardes and José Cunha-Vaz -- Polarization sensitivity / U. Schmidt-Erfurth ... [et al.] -- Adaptive optics in ocular optical coherence tomography / Enrique Josua Fernández and Pablo Artal -- The SL SCAN-1: Fourier domain optical coherence tomography integrated into a slit lamp / F. D. Verbraak and M. Stehouwer. Subjects: Optical coherence tomography. Retina--Tomography. Tomography, Optical Coherence. Diagnostic Techniques, Ophthalmological. Eye Diseases--Diagnosis. Notes: Includes bibliographical references and index. Series: Biological and medical physics, biomedical engineering Biological and medical physics, biomedical engineering. Dewey class no.: 617.7/3507545.

Out of sight, not out of mind: personal and professional perspectives on age-related macular degeneration LCCN: 2011047591: Out of sight, not out of mind: personal and professional perspectives on age-related macular degeneration / Lindy Bergman with the Chicago Lighthouse for People Who Are Blind or Visually Impaired; Jennifer E. Miller, editor. Edition: Updated and expanded version. Published/Created: New York: AFB Press, c2012. Description:

xi, 212 p.: ill.; 26 cm. ISBN: 9780891284857 (pbk.: alk. paper) 9780891284864 (ASCII cd) 9780891284871 (e-book) LC classification: RE661.D3 B47 2012 Related names: Miller, Jennifer E. Chicago Lighthouse (Organization) American Foundation for the Blind. Contents: Gaining a sense of AMD -- Look at it this way -- Keeping your body as healthy as your mind -- Life with less vision, and how to move forward -- Consider the family -- A note to loved ones -- Vision and aging: general and clinical perspectives / Alfred A. Rosenbloom -- Age-related macular degeneration: causes and current medical perspectives / Gerald Allen Fishman -- Emotional and psychological reactions to age-related macular degeneration / David Rakofsky -- The low vision eye exam: what to expect / Kara Crumbliss -- Integrating low vision rehabilitation and assistive technology / Thomas Perski -- The importance of research in low vision / Patricia Grant. Subjects: Macular Degeneration--psychology. Macular Degeneration--etiology. Macular Degeneration--rehabilitation. Notes: Includes bibliographical references (p. 190-206) and index. Dewey class no.: 617.7/35.

Retinopathy: new research LCCN: 2011037394: Retinopathy: new research / editors, Connie Wahler and Lauren Wilkinson. Published/Created: New York: Nova Science Publishers, c2012. Description: xi, 163 p.: ill. (some col.); 26 cm. ISBN: 9781621005858 (hardcover) 1621005852 (hardcover) LC classification: RE551 .R496 2012 Related names: Wahler, Connie. Wilkinson, Lauren. Contents: Nodavirus infection in mari cultures: diagnosis of viral encephalopathy and retinopathy / N. Cherif, H. Azad, S. Hammami -- Prevalence and clinical profile of diabetic retinopathy in India / R. Rajalakshmi ... [et al.] -- Retinopathy of prematurity: historical perspective and new treatment modalities / Kristine Pierce, Timothy Stout -- Isoproterenol therapy improves neuronal function, vascular integrity, and insulin receptor signaling in diabetic retinopathy / Youde Jiang ... [et al.] -- Retinopathies and associated deafness: a bird's eye view / Barkur S. Shastry -- Fight against angiogenesis-related blindness / Dong Hyun Jo, Jeong Hun Kim -- Diabetic retinopathy: relationship with long-term glycemic exposure, glucose variability and cardiovascular risk / J.A. Gimeno-Orna, A.B. Mañas-Martínez -- Image mining approaches for the screening of age-related macular degeneration / Mohd Hanafi Ahmad Hijazi, Frans Coenen, Yalin Zheng -- Diagnosis of retinal anomalies by statistical tests on OCT data / C. Bruni ... [et al.]. Subjects: Retina--Diseases. Retinal Diseases. Notes: Includes bibliographical references and index. Series: Eye and vision research developments Dewey class no.: 617.7/35.

Swept-source optical coherence tomography: a color atlas LCCN: 2015020370: Swept-source optical coherence tomography: a color atlas / Kelvin Yi Chong Teo, Wong Chee Wai, Andrew Shih Hsiang Tsai, Daniel Shu Wei Ting. Published/Produced: New Jersey: World Scientific, 2015. Projected pub date: 1508 Description: p.; cm. ISBN: 9789814704212 (pbk.: alk. paper) 9814704210 (pbk.: alk. paper) LC classification: RE551 Related names: Wong, Chee Wai, author. Tsai, Andrew Shih Hsiang, author. Ting, Daniel Shu Wei, author. Contents: Introduction to SS-OCT -- Retinal vascular disease -- Macula disorders -- Central serous chorioretinopathy -- Age related macular degeneration -- Vitreo macular interface disease -- Myopia -- Inflammatory conditions -- Miscellaneous conditions. Subjects: Retinal Diseases--diagnosis--Atlases. Tomography, Optical Coherence--methods--Atlases. Notes: Includes index. Dewey class no.: 617.7/350757.

Index

B

C

D

E

M

N

O

P

Q

R

S

T

U

V

W

Z